Fluctuation Relations and Nonequilibrium Thermodynamics in Classical and Quantum Systems

Fluctuation Relations and Nonequilibrium Thermodynamics in Classical and Quantum Systems

Editor

Gabriele De Chiara

MDPI • Basel • Beijing • Wuhan • Barcelona • Belgrade • Manchester • Tokyo • Cluj • Tianjin

Editor
Gabriele De Chiara
Centre for Theoretical Atomic,
Molecular and Optical Physics,
Queen's University Belfast
UK

Editorial Office
MDPI
St. Alban-Anlage 66
4052 Basel, Switzerland

This is a reprint of articles from the Special Issue published online in the open access journal *Entropy* (ISSN 1099-4300) (available at: https://www.mdpi.com/journal/entropy/special_issues/Fluctuation_Relations).

For citation purposes, cite each article independently as indicated on the article page online and as indicated below:

LastName, A.A.; LastName, B.B.; LastName, C.C. Article Title. *Journal Name* **Year**, *Article Number*, Page Range.

ISBN 978-3-03936-998-0 (Hbk)
ISBN 978-3-03936-999-7 (PDF)

© 2020 by the authors. Articles in this book are Open Access and distributed under the Creative Commons Attribution (CC BY) license, which allows users to download, copy and build upon published articles, as long as the author and publisher are properly credited, which ensures maximum dissemination and a wider impact of our publications.

The book as a whole is distributed by MDPI under the terms and conditions of the Creative Commons license CC BY-NC-ND.

Contents

About the Editor vii

Preface to "Fluctuation Relations and Nonequilibrium Thermodynamics in Classical and Quantum Systems" ix

Sebastian Deffner
Efficiency of Harmonic Quantum Otto Engines at Maximal Power
Reprinted from: *Entropy* **2018**, *20*, 875, doi:10.3390/e20110875 1

Francisco J. Peña, Oscar Negrete, Gabriel Alvarado Barrios, David Zambrano, Alejandro González, Alvaro S. Nunez, Pedro A. Orellana and Patricio Vargas
Magnetic Otto Engine for an Electron in a Quantum Dot: Classical and Quantum Approach
Reprinted from: *Entropy* **2019**, *21*, 512, doi:10.3390/e21050512 11

Cleverson Cherubim, Frederico Brito and Sebastian Deffner
Non-Thermal Quantum Engine in Transmon Qubits
Reprinted from: *Entropy* **2019**, *21*, 545, doi:10.3390/e21060545 27

Amjad Aljaloud, Sally A. Peyman and Almut Beige
A Quantum Heat Exchanger for Nanotechnology
Reprinted from: *Entropy* **2020**, *22*, 379, doi:10.3390/e22040379 39

Zoë Holmes, Erick Hinds Mingo, Calvin Y.-R. Chen and Florian Mintert
Quantifying Athermality and Quantum Induced Deviations from Classical Fluctuation Relations
Reprinted from: *Entropy* **2020**, *22*, 111, doi:10.3390/e22010111 57

Jader Santos, André Timpanaro and Gabriel Landi
Joint Fluctuation Theorems for Sequential Heat Exchange
Reprinted from: *Entropy* **2020**, *22*, 763, doi:10.3390/e22070763 87

Federico Corberi and Alessandro Sarracino
Probability Distributions with Singularities
Reprinted from: *Entropy* **2019**, *21*, 312, doi:10.3390/e21030312 97

Fabian Bernards, Matthias Kleinmann, Otfried Gühne and Mauro Paternostro
Daemonic Ergotropy: Generalised Measurements and Multipartite Settings
Reprinted from: *Entropy* **2019**, *21*, 771, doi:10.3390/e21080771 117

Onat Arısoy, Steve Campbell and Özgür E. Müstecaplıoğlu
Thermalization of Finite Many-Body Systems by a Collision Model
Reprinted from: *Entropy* **2019**, *21*, 1182, doi:10.3390/e21121182 133

About the Editor

Gabriele De Chiara is a reader in Quantum Information at Queen's University Belfast (UK). He graduated in Palermo (Italy) in 2002, and obtained his PhD at Scuola Normale Superiore in Pisa (Italy) in 2006. Previously, he was a post-doctoral researcher at the BEC Centre in Trento (Italy) and a "Juan de la Cierva" fellow in Barcelona (Spain). He was a founding editor of the community-led journal "Quantum" (2016–2019), and he is currently an associate editor for "Physical Review A" and contributing editor for "Physical Review Research". His main interests include entanglement and quantum correlations in many-body systems, especially critical 1D systems, optical lattices, Coulomb crystals, open quantum systems and, more recently, quantum thermodynamics.

Preface to "Fluctuation Relations and Nonequilibrium Thermodynamics in Classical and Quantum Systems"

Out-of-equilibrium quantum thermodynamics is now establishing itself as a lively and productive area at the intersection of statistical mechanics and quantum information. This success has been spurred on, on one side, by the discovery of classical and quantum fluctuation theorems. On the other side, quantum information theoretic investigations on resource theories and information-powered engines have led to unexpected results. Moreover, advances in experimental quantum technologies have allowed for the demonstrations of thermodynamic devices with small quantum systems.

This Special Issue includes novel results on a diverse range of topics that provide an excellent showcase of the research activities in classical and quantum thermodynamics. In particular, a number of papers present schemes of engines or other thermal devices whose working substance is a small quantum system, e.g. a quantum harmonic oscillator [1], an electron in a quantum dot [2], a transmon qubit [3] and an atomic gas in an optical cavity [4]. Others explore the theory of classical and quantum fluctuation relations [5, 6], including the singular probability distribution of thermodynamic quantities [7]. One contribution investigates the concept of daemonic entropy, arising in schemes of work extraction through generalised quantum measurements [8]. Last but not least, one paper studies the thermalization of many-body systems by a collision model [9]. There are still many open problems, concerning the role of genuine quantum features in thermal devices, the emergence of the laws of thermodynamics from first principles, and the thermalization of closed systems. Moreover, quantum thermodynamics will also play a significant role in the realization of energy-efficient quantum technologies in the near future.

1. Deffner, S. Efficiency of Harmonic Quantum Otto Engines at Maximal Power. Entropy 2018, 20, 875.
2. Peña, F.J.; Negrete, O.; Alvarado Barrios, G.; Zambrano, D.; González, A.; Nunez, A.S.; Orellana, P.A.; Vargas, P. Magnetic Otto Engine for an Electron in a Quantum Dot: Classical and Quantum Approach. Entropy 2019, 21, 512.
3. Cherubim, C.; Brito, F.; Deffner, S. Non-Thermal Quantum Engine in Transmon Qubits. Entropy 2019, 21, 545.
4. Aljaloud, A.; Peyman, S.A.; Beige, A. A Quantum Heat Exchanger for Nanotechnology. Entropy 2020, 22, 379.
5. Holmes, Z.; Hinds Mingo, E.; Chen, C.-R.; Mintert, F. Quantifying Athermality and Quantum Induced Deviations from Classical Fluctuation Relations. Entropy 2020, 22, 111.
6. Santos, J.; Timpanaro, A.; Landi, G. Joint Fluctuation Theorems for Sequential Heat Exchange. Entropy 2020, 22, 763.
7. Corberi, F.; Sarracino, A. Probability Distributions with Singularities. Entropy 2019, 21, 312.
8. Bernards, F.; Kleinmann, M.; Gühne, O.; Paternostro, M. Daemonic Ergotropy: Generalised Measurements and Multipartite Settings. Entropy 2019, 21, 771.
9. Arısoy, O.; Campbell, S.; Müstecaplıoğlu, Ö.E. Thermalization of Finite Many-Body Systems by a Collision Model. Entropy 2019, 21, 1182.

Gabriele De Chiara
Editor

Article

Efficiency of Harmonic Quantum Otto Engines at Maximal Power

Sebastian Deffner

Department of Physics, University of Maryland Baltimore County, Baltimore, MD 21250, USA; deffner@umbc.edu

Received: 24 October 2018; Accepted: 13 November 2018; Published: 15 November 2018

Abstract: Recent experimental breakthroughs produced the first nano heat engines that have the potential to harness quantum resources. An instrumental question is how their performance measures up against the efficiency of classical engines. For single ion engines undergoing quantum Otto cycles it has been found that the efficiency at maximal power is given by the Curzon–Ahlborn efficiency. This is rather remarkable as the Curzon–Alhbron efficiency was originally derived for endoreversible Carnot cycles. Here, we analyze two examples of endoreversible Otto engines within the same conceptual framework as Curzon and Ahlborn's original treatment. We find that for endoreversible Otto cycles in classical harmonic oscillators the efficiency at maximal power is, indeed, given by the Curzon–Ahlborn efficiency. However, we also find that the efficiency of Otto engines made of quantum harmonic oscillators is significantly larger.

Keywords: quantum Otto engine; Curzon–Ahlborn efficiency; endoreversible quantum thermodynamics

1. Introdcution

It is a standard exercise of thermodynamics to compute the efficiency of engines, i.e., to determine the relative work output for devices undergoing cyclic transformations on the thermodynamic manifold [1]. Like few other applications the study of heat engines illustrates the versatility of thermodynamic concepts, since universally valid bounds can be obtained purely from macroscopic, phenomenological knowledge about physical systems. However, all ideal cycles, such as the Carnot, Stirling, Otto, Diesel, etc. cycles are only of limited practical importance, as they are comprised of quasistatic, infinitely slow state transformations. Therefore, the power output of an ideal engine is strictly zero [1].

All real engines operate in finite time, and thus their working medium is almost never in equilibrium with the environment. Moreover, a more practical question is to determine the efficiency at maximal power output, rather than focusing only at the ideal, maximal efficiency (at zero power). In a seminal paper [2], Curzon and Ahlborn tackled this problem within the framework of *endoreversible thermodynamics* [3].

At the core of endoreversible thermodynamics is the idea of *local equilibrium*: Imagine an engine, whose working medium is in a state of thermal equilibrium of temperature T. However, T is not equal to the temperature of the environment, T_{bath}, and thus there is a temperature gradient at the boundaries of the engine. One then studies the engine as it slowly undergoes a cyclic state transformation, where slow means that the working medium remains *locally* in equilibrium at all times. However, since the cycle does operate in finite time, the working medium never fully equilibrates with the environment. Therefore, from the point of view of the environment the device undergoes an irreversible cycle. Such state transformations are called *endoreversible* [3], which means that locally the transformation is reversible, but globally irreversible.

Curzon and Ahlborn showed [2] that the efficiency of a Carnot engine undergoing an endoreversible cycle at maximal power is given by,

$$\eta_{CA} = 1 - \sqrt{\frac{T_c}{T_h}}, \tag{1}$$

where T_c and T_h are the temperatures of the cold and hot reservoirs, respectively. Remarkably, it has been found that η_{CA} (1) is also assumed by many, physically different engines at maximal power, such as an endoreversible Otto engine with an ideal gas as working medium [4], the endoreversible Stirling cycle [5], Otto engines in open quantum systems in the quasistatic limit [6], or a single ion in a harmonic trap undergoing a quantum Otto cycle [7,8]. On the other hand, it also has been shown that whether or not a finite time Carnot cycle assumes η_{CA} is determined by the "symmetry" of dissipation [9], and the efficiency of an Otto engine working with a single Brownian particle in a harmonic trap is determined by the specific parameterization of the trap's stiffness [10].

In particular, the recent experimental breakthroughs in the implementation of nanosized heat engines [11,12] that could principally exploit quantum resources [13–24] pose the question whether their behavior can be universally characterized. For instance, Reference [6] suggested that to describe the efficiency at maximal power η_{CA} could be such a universal result, at least for a class of engines. However, the Curzon–Ahlborn efficiency (1) was originally derived for endoreversible Carnot cycles, which is independent on the nature of the working medium. On the other hand, a standard textbook exercise shows that the Otto efficiency is dependent on the equation of state, i.e., on the specific working medium [1]. Therefore, it would actually be more natural to expect that the efficiency at maximal power strongly depends on the nature of working medium. Similar conclusions have been drawn, for instance, in the thermodynamic analysis of photovolatic cells [25–27].

In addition, the quantum Otto cycle is typically comprised of two thermalization and two unitary strokes [28–30]. For cycles involving only unitary strokes [7,8] the assumption of local equilibrium is almost never justified, and thus it becomes even more remarkable that at maximal power output a quantum Otto cycle in a parametric, harmonic oscillator operates with the Curzon–Ahlborn efficiency [7,8]. Also see Reference [6] for a more detailed treatment from open quantum dynamics. Therefore, the question arises whether this is a peculiarity of the quantum Otto cycle in the harmonic oscillator, or whether there is something more fundamental and universal about η_{CA}.

The purpose of the present work is to revisit these longstanding questions and study the endoreversible Otto cycle in a conceptually simple and pedagogical approach similar to Curzon and Ahlborn's original treatment [2]. To this end, we compute the efficiency at maximal power for two examples of endoreversible Otto engines. We start with a classical version, for which the working medium is a single Brownian particle in a harmonic trap. Maximizing the power output with respect to the compression ratio, we find analytically that the efficiency is indeed given by η_{CA} (1). As a second example we study a quantum engine, whose working medium is a quantum harmonic oscillator ultraweakly coupled to the thermal environment. We find that in this case the efficiency is larger than η_{CA} (1), which demonstrates that the Curzon–Ahlborn efficiency is *not* universal at maximal power. An advantage of the present treatment is that it is somewhat more pedagogical than earlier works on the topic. The present derivation is entirely based on the phenomenological framework of endoreversible thermodynamics. Thus, e.g., neither the full quantum dynamics [6] nor the linear response problem [10] have to be solved.

2. Carnot Engine at Maximal Power

We begin by briefly reviewing the main gist of Reference [2] and by establishing notions and notation. In particular, we focus on the limits and assumptions that lead to the Curzon–Ahlborn efficiency (1) for endoreversible Carnot engines.

The ideal Carnot cycle consists of two isothermal processes during which the systems absorbs/exhausts heat and two thermodynamically adiabatic, i.e., isentropic strokes [1]. During the isentropic strokes the working medium does not exchange heat with the thermal reservoirs, and thus its state can be considered to be independent of the environment. Therefore, we only have to modify the treatment of the isothermal strokes during which the working medium will be in a local equilibrium state at different temperature than the temperature of the hot and cold reservoir, respectively.

In particular, during the hot isotherm the working medium is assumed to be a little cooler than the hot environment at T_h. Thus, during the whole stroke the system absorbs the heat

$$Q_h = \lambda_h \tau_h (T_h - T_{hw}), \tag{2}$$

where τ_h is the stroke time, $T_{h,w}$ is the temperature of the working medium, and λ_h is a constant depending on thickness and thermal conductivity of the boundary separating working medium and environment. Note that Equation (2) is nothing else but a discretized version of Fourier's law for heat conduction [1]. We will see shortly that for Otto cycles the rate of heat flux can no longer be assumed to be constant, since we need to account for the change in temperature during the isochoric strokes.

Similarly, during the cold isotherm the system is a little warmer than the cold reservoir at T_c. Hence, the exhausted heat can be written as

$$Q_c = \lambda_c \tau_c (T_{cw} - T_c) \tag{3}$$

where λ_c is the cold heat transfer coefficient.

As mentioned above, the adiabatic strokes are unmodified, but note that the cycle is taken to be reversible with respect to the *local temperatures* of the working medium. Hence, we can write

$$\Delta S_h = -\Delta S_c \quad \text{and thus} \quad \frac{Q_h}{T_{hw}} = \frac{Q_c}{T_{cw}}. \tag{4}$$

Equation (4) allows to relate the stroke times τ_h and τ_c to the heat transfer coefficients λ_h and λ_c. We are now interested in determining the efficiency at maximal power. To this end, we write the power output of the cycle as

$$P(\delta T_h, \delta T_c) = \frac{Q_h - Q_c}{\gamma(\tau_h + \tau_c)} \tag{5}$$

where $\delta T_h = T_h - T_{hw}$ and $\delta T_c = T_{cw} - T_c$. In Equation (5) we introduced the total cycle time $\tau_{cyc} = \gamma(\tau_h + \tau_c)$, and thus $\gamma \equiv \tau_{cyc}/(\tau_h + \tau_c)$. Note that this neglects any explicit dependence of the analysis on the lengths of the adiabatic strokes. We exclusively focus on the isotherms, i.e, on the temperature differences between working medium and the hot and cold reservoirs.

It is worth emphasizing that in the present problem we have four free parameters, namely hot and cold temperatures of the working substance, T_{hw} and T_{cw}, and the stroke times τ_h and τ_c. The balance equation for the entropy (4) allows to eliminate two of these, and Curzon and Ahlborn chose to eliminate τ_h and τ_c [2].

Thus, we maximize the power $P(\delta T_h, \delta T_c)$ as a function of the difference in temperatures between working substance and environment. After a few lines of algebra one obtains [2],

$$P_{\max} = \frac{\lambda_h \lambda_c}{\gamma} \left(\frac{\sqrt{T_h} - \sqrt{T_c}}{\sqrt{\lambda_h} + \sqrt{\lambda_c}} \right)^2, \tag{6}$$

where the maximum is assumed for

$$\frac{\delta T_h}{T_h} = \frac{1 - \sqrt{T_c/T_h}}{1 + \sqrt{\lambda_h/\lambda_c}} \quad \text{and} \quad \frac{\delta T_c}{T_c} = \frac{\sqrt{T_h/T_c} - 1}{1 + \sqrt{\lambda_c/\lambda_h}} \tag{7}$$

From these expressions we can now compute the efficiency. We have,

$$\eta = \frac{Q_h - Q_c}{Q_h} = 1 - \frac{T_{cw}}{T_{hw}} = 1 - \frac{T_c + \delta T_c}{T_h - \delta T_h} \qquad (8)$$

where we used Equation (4). Thus, the efficiency of an endoreversible Carnot cycle at maximal power output becomes

$$\eta_{CA} = 1 - \sqrt{\frac{T_c}{T_h}}, \qquad (9)$$

which only depends on the temperatures of the hot and cold reservoirs.

In the following, we will apply exactly the same reasoning to the endoreversible Otto cycle.

3. Endoreversible Otto Cycle

The standard Otto cycle is a four-stroke cycle consisting of isentropic compression, isochoric heating, isentropic expansion, and ischoric cooling [1]. Thus, we have in the endoreversible regime:

3.1. Isentropic Compression

During the isentropic strokes the working substance does not exchange heat with the environment. Therefore, the thermodynamic state of the working substance can be considered independent of the environment, and the endoreversible description is identical to the equilibrium cycle. From the first law of thermodynamics, $\Delta E = Q + W$, we have,

$$Q_{\text{comp}} = 0 \quad \text{and} \quad W_{\text{comp}} = E(T_2, \omega_2) - E(T_1, \omega_1) \qquad (10)$$

where Q_{comp} is the heat exchanged, and W_{comp} is the work performed during the compression. Moreover, ω denotes the work parameter, such as the inverse volume of a piston or the frequency of a harmonic oscillator (20).

3.2. Isochoric Heating

During the isochoric strokes the work parameter is held constant, and the system exchanges heat with the environment. Thus, we have for isochoric heating

$$Q_h = E(T_3, \omega_2) - E(T_2, \omega_2) \quad \text{and} \quad W_h = 0. \qquad (11)$$

In complete analogy to Curzon and Ahlborn's original analysis [2] we now assume that the working substance is in a state of local equilibrium, but also that the working substance never fully equilibrates with the hot reservoir. Therefore, we can write

$$T(0) = T_2 \quad \text{and} \quad T(\tau_h) = T_3 \quad \text{with} \quad T_2 < T_3 \leq T_h, \qquad (12)$$

where as before τ_h is the duration of the stroke.

Note that in contrast to the Carnot cycle the Otto cycle does not involve isothermal strokes, and, hence, the rate of heat flux is not constant. Rather, we have to explicitly account for the change in temperature from T_2 to T_3. To this end, Equation (2) is replaced by Fourier's law [1],

$$\frac{dT}{dt} = -\alpha_h \left(T(t) - T_h \right) \qquad (13)$$

where α_h is a constant depending on the heat conductivity and heat capacity of the working substance.

Equation (13) can be solved exactly, and we obtain the relation

$$T_3 - T_h = (T_2 - T_h) \exp\left(-\alpha_h \tau_h\right). \qquad (14)$$

In the following, we will see that Equation (14) is instrumental in reducing the number of free parameters.

3.3. Isentropic Expansion

In complete analogy to the compression, we have for the isentropic expansion,

$$Q_{\exp} = 0 \quad \text{and} \quad W_{\exp} = E(T_4, \omega_1) - E(T_3, \omega_2). \tag{15}$$

3.4. Isochoric Cooling

Heat and work during the isochoric cooling read,

$$Q_c = E(T_1, \omega_1) - E(T_4, \omega_1) \quad \text{and} \quad W_c = 0, \tag{16}$$

where we now have

$$T(0) = T_4 \quad \text{and} \quad T(\tau_c) = T_1 \quad \text{with} \quad T_4 > T_1 \geq T_c. \tag{17}$$

Similarly to above (13) the heat transfer is described by Fourier's law

$$\frac{dT}{dt} = -\alpha_c \left(T(t) - T_c \right), \tag{18}$$

where α_c is a constant characteristic for the cold stroke. From the solution of Equation (18) we now obtain

$$T_1 - T_c = (T_4 - T_c) \exp(-\alpha_c \tau_c), \tag{19}$$

which properly describes the decrease in temperature from T_4 back to T_1.

4. Classical Harmonic Engine

To continue the analysis we now need to specify the internal energy E. As a first example, we consider a classical Brownian particle trapped in a harmonic oscillator. The bare Hamiltonian reads,

$$H(p, x) = \frac{p^2}{2m} + \frac{1}{2} m \omega^2 x^2, \tag{20}$$

where m is the mass of the particle.

For a particle in thermal equilibrium the Gibbs entropy, S, and the internal energy, E, are

$$\frac{S}{k_B} = 1 + \ln\left(\frac{k_B T}{\hbar \omega}\right) \quad \text{and} \quad E = k_B T, \tag{21}$$

where we introduced Boltzmann's constant, k_B.

Note, that from Equation (21) we obtain a relation between the frequencies, ω_1 and ω_2 and the four temperatures, T_1, T_2, T_3, and T_4. To this end, consider the isentropic strokes, for which we have

$$S(T_2, \omega_2) = S(T_1, \omega_1) \quad \text{and} \quad S(T_4, \omega_1) = S(T_3, \omega_2), \tag{22}$$

which is fulfilled by

$$T_1 \omega_2 = T_2 \omega_1 \quad \text{and} \quad T_3 \omega_1 = T_4 \omega_2. \tag{23}$$

We are now equipped with all the ingredients necessary to compute the endoreversible efficiency,

$$\eta = -\frac{W_{\text{tot}}}{Q_h}. \tag{24}$$

In complete analogy to fully reversible cycles [1], Equation (24) can be written as

$$\eta = 1 - \frac{T_4 - T_1}{T_3 - T_2}, \tag{25}$$

where we used the explicit from of the internal energy E (21). Further, using Equations (23) the endoreversible Otto efficiency becomes

$$\eta = 1 - \frac{\omega_1}{\omega_2} \equiv 1 - \kappa, \tag{26}$$

which defines the compression ratio, κ. Observe that the endoreversible efficiency takes the same form as its reversible counter part [1]. However, in Equation (25) the temperatures correspond the local equilibrium state of the working substance, and not to a global equilibrium with the environment.

Similarly to Curzon and Ahlborn's treatment of the endoreversible Carnot cycle [2] we now compute the efficiency for a value of κ, at which the power (5) is maximal. We begin by re-writing the total work with the help of the compression ratio κ and Equations (23) as,

$$W_{\text{tot}} = W_{\text{comp}} + W_{\text{exp}} = (\kappa - 1) \, k_B \, (T_2 - T_3). \tag{27}$$

Further using Equation (14) we obtain

$$W_{\text{tot}} = (\kappa - 1)(1 - \exp(-\alpha_h \tau_h)) \, k_B \, (T_2 - T_h), \tag{28}$$

which only depends on the free parameters T_2, κ, and τ_h. Of these three, we can eliminate one more, by combing Equations (14) and (19), and we have

$$T_2 = \frac{T_c \left(e^{\alpha_c \tau_c} - 1\right) + \kappa T_h \left(1 - e^{-\alpha_h \tau_h}\right)}{\kappa \left(e^{\alpha_c \tau_c} - e^{-\alpha_h \tau_h}\right)}. \tag{29}$$

Finally, the power output (5) takes the form,

$$P = \frac{2(\kappa - 1) \, k_B \, (T_c - \kappa T_h)}{\gamma \kappa (\tau_c + \tau_h)} \, \frac{\sinh(\alpha_c \tau_c / 2) \sinh(\alpha_h \tau_h / 2)}{\sinh[(\alpha_c \tau_c + \alpha_h \tau_h)/2]}. \tag{30}$$

Remarkably the power output, $P(\kappa, \tau_h, \tau_c)$, factorizes into a contribution that only depends on the compression ratio, κ, and another term that is governed by the stroke times, τ_c and τ_h,

$$P(\kappa, \tau_h, \tau_c) = f_1(\kappa) f_2(\tau_h, \tau_c). \tag{31}$$

It is then a simple exercise to show that $P(\kappa, \tau_h, \tau_c)$ is maximized for any value of τ_h and τ_c if we have,

$$P_{\max} = P(\kappa_{\max}) \quad \text{with} \quad \kappa_{\max} = \sqrt{\frac{T_c}{T_h}}. \tag{32}$$

Therefore, the efficiency at maximal power reads,

$$\eta = 1 - \sqrt{\frac{T_c}{T_h}}. \tag{33}$$

In conclusion, we have shown that for the classical harmonic oscillator the efficiency at maximal power of an endoreversible Otto cycle (24) is indeed given by the Curzon–Ahlborn efficiency (1).

It is worth emphasizing that for the endoreversible Otto cycle we started with six free parameters, the four temperatures T_1, T_2, T_3, and T_4, and the two stroke times, τ_h and τ_c. Of these, we succeeded in eliminating three, by explicitly using Fourier's law for the heat transfer, Equations (13) and (18), and the explicit expressions for the entropy and the internal energy (21). Therefore, one would not expect to obtain the same result (33) for other working substances such as the quantum harmonic oscillator.

5. Quantum Harmonic Engine

For the remainder of this analysis we will be interested in a quantum harmonic oscillator in the ultraweak coupling limit [31]. In this limit, a "small" quantum system interacts only weakly with a large Markovian heat bath, such that the stationary state is given by a thermal equilibrium distribution. This situation is similar to the model studied in Reference [6], however in the present case we will not have to solve the full quantum dynamics.

The equilibrium state is given by a Gibbs state, $\rho \propto \exp(-H/k_B T)$, where ρ is the density operator. Accordingly, the internal energy reads

$$E = \frac{\hbar \omega}{2} \coth\left(\frac{\hbar \omega}{2 k_B T}\right) \tag{34}$$

and the entropy becomes

$$\frac{S}{k_B} = \frac{\hbar \omega}{2 k_B T} \coth\left(\frac{\hbar \omega}{2 k_B T}\right) - \ln\left[\frac{1}{2}\sinh\left(\frac{\hbar \omega}{2 k_B T}\right)\right]. \tag{35}$$

Despite the functional form of S being more involved, we notice that the four temperatures and the two frequencies are still related by the same Equation (23). Thus, it can be shown [6] that the efficiency of an endoreversible Otto cycle in a quantum harmonic oscillators also reads,

$$\eta = 1 - \kappa. \tag{36}$$

Following the analogous steps that led to Equation (30) we obtain for the power output of an endoreversible quantum Otto engine,

$$P = \operatorname{csch}\left[\frac{\hbar \omega_2 \kappa}{2} \frac{e^{\alpha_c T_c + \alpha_h T_h} - 1}{T_c (e^{\alpha_c T_c} - 1) + \kappa T_h e^{\alpha_c T_c}(e^{\alpha_h T_h} - 1)}\right] \operatorname{csch}\left[\frac{\hbar \omega_2 \kappa}{2} \frac{e^{\alpha_c T_c + \alpha_h T_h} - 1}{T_c e^{\alpha_h T_h}(e^{\alpha_c T_c} - 1) + \kappa T_h (e^{\alpha_h T_h} - 1)}\right] \\ \times \frac{\hbar \omega_2}{2} \frac{1-\kappa}{T_c + T_h} \sinh\left[\frac{\hbar \omega_2 \kappa}{2} \frac{(\kappa T_h - T_c)(e^{\alpha_c T_c + \alpha_h T_h} - 1)(e^{\alpha_h T_h} - 1)(e^{\alpha_c T_c} - 1)}{(T_c(e^{\alpha_c T_c} - 1) + \kappa T_h e^{\alpha_c T_c}(e^{\alpha_h T_h} - 1))(T_c e^{\alpha_h T_h}(e^{\alpha_c T_c} - 1) + \kappa T_h (e^{\alpha_h T_h} - 1))}\right] \tag{37}$$

where we set $k_B = 1$. We immediately observe that in contrast to the classical case (30) the expression no longer factorizes. Consequently, the value of κ, for which P is maximal does depend on the stroke times τ_h and τ_c.

Due to the somewhat cumbersome expression (37) we chose to find the maximum of $P(\kappa, \tau_h, \tau_c)$ numerically. In Figure 1 we illustrate our findings in the high temperature limit, $\hbar \omega_2 / k_B T_c \ll 1$. Consistently with our classical example, the efficiency is given by Equation (33), which was also found in Reference [6] for quasistatic cycles. It is worth emphasizing that Figure 1 was obtained numerically for a specific choice of parameters. However, the above, classical analysis revealed that in the limit of high temperatures the result, namely that the efficiency at maximal power is given by the Curzon–Ahlborn efficiency (33), becomes independent of all parameters but the temperatures of the hot and cold reservoirs.

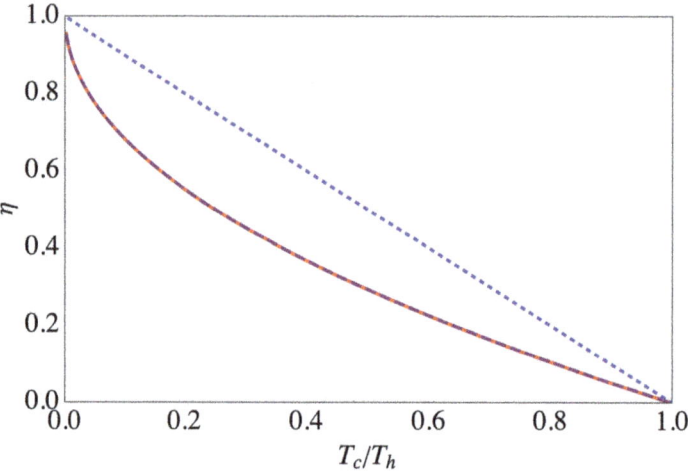

Figure 1. Efficiency of the endoreversible Otto cycle at maximal power (red, solid line), together with the Curzon–Ahlborn efficiency (purple, dashed line) and the Carnot efficiency (blue, dotted line) in the high temperature limit, $\hbar\omega_2/k_B T_c = 0.1$. Other parameters are $\alpha_c = 1$, $\alpha_h = 1$, and $\gamma = 1$.

Figure 2 depicts the efficiency at maximal power (36) as a function of T_c/T_h in the deep quantum regime, $\hbar\omega_2/k_B T_c \gg 1$. In this case, we find that the quantum efficiency is larger than the Curzon–Ahlborn efficiency (33). From a thermodynamics' point-of-view this finding is not really surprising since already in reversible cycles the efficiency strongly depends on the equation of state.

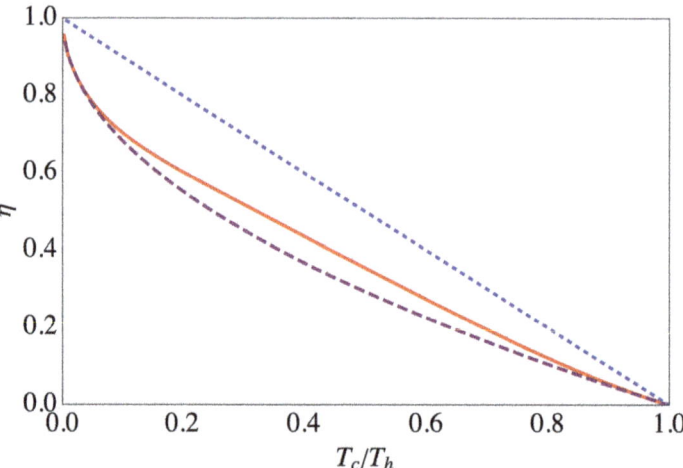

Figure 2. Efficiency of the endoreversible Otto cycle at maximal power (red, solid line), together with the Curzon–Ahlborn efficiency (purple, dashed line) and the Carnot efficiency (blue, dotted line) in the deep quantum regime, $\hbar\omega_2/k_B T_c = 10$. Other parameters are $\alpha_c = 1$, $\alpha_h = 1$, and $\gamma = 1$.

In conclusion, we have shown explicitly that contrary to anecdotal evidence in the literature [4,6–8,12] the efficiency at maximal power is *not* universally given by the Curzon–Ahlborn efficiency—not even for the harmonic oscillator. The natural question now is if and how this "quantum supremacy" can be exploited in the design and experimental implementation of nano engines. This, however, we leave for future work.

6. Concluding Remarks

In the present work we have computed the efficiency at maximal power for two examples of the endoreversible Otto engine. We have found that in the case of a classical harmonic oscillator the efficiency is identical to the Curzon–Ahlborn expression originally found for endoreversible Carnot cycles. However, we have also shown that for engines operating with quantum harmonic oscillators the efficiency significantly differs from the classical expression. These findings are consistent with References [6,10], where it was argued that the efficiency should be governed by internal friction and specific driving protocols, respectively. The advantage of the present analysis is, however, that our results were obtained entirely from the phenomenological equations of endoreversible thermodynamics. Neither the quantum master equation [6] nor the linear response problem [10] had to be solved explicitly.

Finally, we note that the present conclusions are a consequence of the differing equations of state for the classical and quantum harmonic oscillator. More precisely, the maximal power output is governed by the different expressions for the internal energies. As such, the conclusions drawn in this work are more "thermodynamical" as they are "quantum". By this we mean, that it is entirely possible to find classical working substances, for which the efficiency at maximal power is not given by the Curzon–Ahlborn efficiency. We also have not excluded the existence of other quantum working substance, for which are described by the Curzon–Ahlborn efficiency. However, the hunt for these systems we also leave for future work.

Funding: S.D. acknowledges support from the U.S. National Science Foundation under Grant No. CHE-1648973.

Acknowledgments: It is a pleasure to thank Gregory Huxtable for enjoyable discussions during an early stage of this project, and Steve Campbell, Obinna Abah, and Marcus V. S. Bonança for many years of fruitful exchange of ideas.

Conflicts of Interest: The author declares no conflict of interest.

References

1. Callen, H. *Thermodynamics and an Introduction to Thermostastistics*; Wiley: New York, NY, USA, 1985.
2. Curzon, F.L.; Ahlborn, B. Efficiency of a Carnot engine at maximum power output. *Am. J. Phys.* **1975**, *43*, 22–24. [CrossRef]
3. Hoffmann, K.H.; Burzler, J.M.; Schubert, S. Endoreversible thermodynamics. *J. Non-Equilib. Thermodyn.* **1997**, *22*, 311–355. [CrossRef]
4. Leff, H.S. Thermal efficiency at maximum work output: New results for old heat engines. *Am. J. Phys.* **1987**, *55*, 602–610. [CrossRef]
5. Erbay, L.B.; Yavuz, H. Analysis of the Stirling heat engine at maximum power conditions. *Energy* **1997**, *22*, 645–650. [CrossRef]
6. Rezek, Y.; Kosloff, R. Irreversible performance of a quantum harmonic heat engine. *New J. Phys.* **2006**, *8*, 83. [CrossRef]
7. Abah, O.; Roßnagel, J.; Jacob, G.; Deffner, S.; Schmidt-Kaler, F.; Singer, K.; Lutz, E. Single-Ion Heat Engine at Maximum Power. *Phys. Rev. Lett.* **2012**, *109*, 203006. [CrossRef] [PubMed]
8. Roßnagel, J.; Abah, O.; Schmidt-Kaler, F.; Singer, K.; Lutz, E. Nanoscale Heat Engine Beyond the Carnot Limit. *Phys. Rev. Lett.* **2014**, *112*, 030602. [CrossRef] [PubMed]
9. Esposito, M.; Kawai, R.; Lindenberg, K.; Van den Broeck, C. Efficiency at Maximum Power of Low-Dissipation Carnot Engines. *Phys. Rev. Lett.* **2010**, *105*, 150603. [CrossRef] [PubMed]
10. Bonança, M.V.S. Approaching Carnot efficiency at maximum power in linear response regime. *arXiv* **2018**, arXiv:1809.09163.
11. Roßnagel, J.; Dawkins, S.T.; Tolazzi, K.N.; Abah, O.; Lutz, E.; Schmidt-Kaler, F.; Singer, K. A single-atom heat engine. *Science* **2016**, *352*, 325–329. [CrossRef] [PubMed]
12. Klaers, J.; Faelt, S.; Imamoglu, A.; Togan, E. Squeezed Thermal Reservoirs as a Resource for a Nanomechanical Engine beyond the Carnot Limit. *Phys. Rev. X* **2017**, *7*, 031044. [CrossRef]

13. Scovil, H.E.D.; Schulz-DuBois, E.O. Three-Level Masers as Heat Engines. *Phys. Rev. Lett.* **1959**, *2*, 262. [CrossRef]
14. Scully, M.O. Quantum Afterburner: Improving the Efficiency of an Ideal Heat Engine. *Phys. Rev. Lett.* **2002**, *88*, 050602. [CrossRef] [PubMed]
15. Scully, M.O.; Zubairy, M.S.; Agarwal, G.S.; Walther, H. Extracting Work from a Single Heat Bath via Vanishing Quantum Coherence. *Science* **2003**, *299*, 862–864. [CrossRef] [PubMed]
16. Scully, M.O.; Chapin, K.R.; Dorfman, K.E.; Kim, M.B.; Svidzinsky, A. Quantum heat engine power can be increased by noise-induced coherence. *Proc. Natl. Acad. Sci. USA* **2011**, *108*, 15097–15100. [CrossRef] [PubMed]
17. Zhang, K.; Bariani, F.; Meystre, P. Quantum Optomechanical Heat Engine. *Phys. Rev. Lett.* **2014**, *112*, 150602. [CrossRef] [PubMed]
18. Gardas, B.; Deffner, S. Thermodynamic universality of quantum Carnot engines. *Phys. Rev. E* **2015**, *92*, 042126. [CrossRef] [PubMed]
19. Hardal, A.Ü.C.; Müstecaplıoğlu, Ö.E. Superradiant Quantum Heat Engine. *Sci. Rep.* **2015**, *5*, 12953. [CrossRef] [PubMed]
20. Cavina, V.; Mari, A.; Giovannetti, V. Slow Dynamics and Thermodynamics of Open Quantum Systems. *Phys. Rev. Lett.* **2017**, *119*, 050601. [CrossRef] [PubMed]
21. Roulet, A.; Nimmrichter, S.; Taylor, J.M. An autonomous single-piston engine with a quantum rotor. *Quantum Sci. Technol.* **2018**, *3*, 035008. [CrossRef]
22. Cherubim, C.; Brito, F.; Deffner, S. Non-thermal quantum engine in transmon qubits. *arXiv* **2018**, arXiv:1810.04226
23. Niedenzu, W.; Mukherjee, V.; Ghosh, A.; Kofman, A.G.; Kurizki, G. Quantum engine efficiency bound beyond the second law of thermodynamics. *Nat. Commun.* **2018**, *9*, 165. [CrossRef] [PubMed]
24. Ronzani, A.; Karimi, B.; Senior, J.; Chang, Y.C.; Peltonen, J.T.; Chen, C.; Pekola, J.P. Tunable photonic heat transport in a quantum heat valve. *Nat. Phys.* **2018**, *14*, 991. [CrossRef]
25. Scully, M.O. Quantum Photocell: Using Quantum Coherence to Reduce Radiative Recombination and Increase Efficiency. *Phys. Rev. Lett.* **2010**, *104*, 207701. [CrossRef] [PubMed]
26. Dorfman, K.E.; Svidzinsky, A.A.; Scully, M.O. Increasing Photovoltaic Power by Noise Induced Coherence Between Intermediate Band States. *Coherent Opt. Phenom.* **2013**, *1*, 42–49. [CrossRef]
27. Einax, M.; Nitzan, A. Network Analysis of Photovoltaic Energy Conversion. *J. Phys. Chem. C* **2014**, *118*, 27226–27234. [CrossRef]
28. Quan, H.T.; Liu, Y.X.; Sun, C.P.; Nori, F. Quantum thermodynamic cycles and quantum heat engines. *Phys. Rev. E* **2007**, *76*, 031105. [CrossRef] [PubMed]
29. Kosloff, R. Quantum Thermodynamics: A Dynamical Viewpoint. *Entropy* **2013**, *15*, 2100–2128. [CrossRef]
30. Kosloff, R.; Rezek, Y. The Quantum Harmonic Otto Cycle. *Entropy* **2017**, *19*, 136. [CrossRef]
31. Spohn, H.; Lebowitz, J.L. Irreversible Thermodynamics for Quantum Systems Weakly Coupled to Thermal Reservoirs. *Adv. Chem. Phys.* **1978**, *38*, 109–142. [CrossRef]

© 2018 by the author. Licensee MDPI, Basel, Switzerland. This article is an open access article distributed under the terms and conditions of the Creative Commons Attribution (CC BY) license (http://creativecommons.org/licenses/by/4.0/).

Article

Magnetic Otto Engine for an Electron in a Quantum Dot: Classical and Quantum Approach

Francisco J. Peña [1,*], **Oscar Negrete** [1,2], **Gabriel Alvarado Barrios** [2,3], **David Zambrano** [1], **Alejandro González** [1], **Alvaro S. Nunez** [3,4], **Pedro A. Orellana** [1] and **Patricio Vargas** [1,3]

1. Departamento de Física, Universidad Técnica Federico Santa María, Casilla 110-V, 2390123 Valparaíso, Chile; oscar.negrete@usm.cl (O.N.); david.zambrano@usm.cl (D.Z.); alejandro.gonzalezi@usm.cl (A.G.); pedro.orellana@usm.cl (P.A.O.); patricio.vargas@usm.com (P.V.)
2. Departamento de Física, Universidad de Santiago de Chile (USACH), Avenida Ecuador 3493, 9170022 Santiago, Chile; gabriel.alvarado@usach.cl
3. Centro para el Desarrollo de la Nanociencia y la Nanotecnología, 8320000 Santiago, Chile; alnunez@dfi.uchile.cl
4. Departamento de Física, Facultad de Ciencias Físicas y Matemáticas, Universidad de Chile, Casilla 487-3, 8370456 Santiago, Chile
* Correspondence: f.penarecabarren@gmail.com or francisco.penar@usm.cl

Received: 7 January 2019; Accepted: 1 March 2019; Published: 20 May 2019

Abstract: We studied the performance of classical and quantum magnetic Otto cycle with a working substance composed of a single quantum dot using the Fock–Darwin model with the inclusion of the Zeeman interaction. Modulating an external/perpendicular magnetic field, in the classical approach, we found an oscillating behavior in the total work extracted that was not present in the quantum formulation. We found that, in the classical approach, the engine yielded a greater performance in terms of total work extracted and efficiency than when compared with the quantum approach. This is because, in the classical case, the working substance can be in thermal equilibrium at each point of the cycle, which maximizes the energy extracted in the adiabatic strokes.

Keywords: magnetic cycle; quantum otto cycle; quantum thermodynamics

1. Introduction

The study of quantum heat engines (QHEs) [1] is focused on the search and design of efficient nanoscale devices operating with a quantum working substance. These devices are characterized by their working substance, the thermodynamic cycle of operation, and the dynamics that govern the cycle [2–26]. Among the cycles in which the engine may operate, the Carnot and Otto cycles have received increasing attention. In particular, the quantum Otto cycle has been considered for various working substances such as spin-1/2 systems [27,28] and harmonic oscillators [29], among others. Recently, an increasing number of experimental realizations for the quantum Otto cycle has been proposed in the literature [30–33]. Furthermore, it has been shown that thermal machines can be reduced to the limits of single atoms [34].

Previous studies of the quantum Otto cycle embedding working substances with magnetic properties have highlighted the role of degeneracy in the energy spectrum on the performance of the engine [35–41]. In this same framework, we highlight the work of Mehta and Johal [38], who studied a quantum Otto engine in the presence of level degeneracy, finding an enhancement of work and efficiency for two-level particles with a degeneracy in the excited state. In addition, Azimi et al. presented the study of a quantum Otto engine operating with a working substance of a single phase multiferroic $LiCu_2O_2$ tunable by external electromagnetic fields [39], which was extended by

Chotorlishvili et al. [40] under the implementation of shortcuts to adiabaticity, finding an optimal output power for the proposed machine.

On the other hand, the classical description of the Otto cycle is characterized by state variables that are well-defined at each point of the cycle. In this sense, the main difference between the classical and quantum approach is that in the classical cycle the working substance can be at thermal equilibrium after each stroke. Classically, the adiabatic strokes are determined by the isentropic condition, which allows determining the state variables. For many systems, such as diamagnetic systems, which were considered in this study, the relation between the thermodynamics variables involved in the adiabatic stroke is not trivial in general and must be solved numerically [41].

In particular, it is interesting to compare the classical and quantum approaches for the same working substance and establish the conditions for each case appropriately. In this framework, several recent studies have focused on employing quantum coherence in the working fluid for enhancing the performance of the engine [42–44]. Recently, an interesting regime called "sudden cycles" [45] has been explored in an incoherent formulation avoiding off-diagonal elements of the density matrix, characterized by finite cooling power [46].

In this work, we study the classical and quantum performance of a multi-level Otto cycle in a diagonal formulation of the density matrix operator, where the working substance comprises a nanosized quantum dot under a controllable external magnetic field. This system is described by the Fock–Darwin model [47,48] that represents an accurate model for a semiconductor quantum dot. For this diamagnetic system, we find the point at which the quantum total work extracted becomes smaller than the classical one and we report, in the classical approach, an oscillating behaviour in the total work extracted that is not perceptible under the quantum formulation.

2. Model

Let us consider a system given by an electron in the presence of a parabolic potential and external magnetic field **B**. The Hamiltonian that describes the system is given by

$$\hat{\mathcal{H}} = \frac{1}{2m^*}(\mathbf{p} + e\mathbf{A})^2 + U_D(x,y), \tag{1}$$

where m^* is the effective electron mass, **A** is the total vector potential, and the term $U_D(x,y)$ is given by

$$U_D(x,y) = \frac{1}{2}m^*\omega_0^2\left(x^2 + y^2\right), \tag{2}$$

which corresponds to an attractive potential describing the effect of the dot on the electron. The quantity ω_0 is the parabolic trap frequency and can be controlled geometrically. If we consider a constant perpendicular magnetic field in the form

$$\mathbf{B} = B\hat{z}, \tag{3}$$

and the use of the vector potential **A** in the symmetric gauge (i.e., $\mathbf{A} = \frac{B}{2}(-y, x, 0)$), the solution of the eigenvalues of the Schrödinger equation are given by

$$E_{nm} = \hbar\Omega\left(2n + |m| + 1\right) + \frac{1}{2}\hbar\omega_c m. \tag{4}$$

where $\omega_c = \frac{eB}{m^*}$ is the cyclotron frequency, and n and m are the radial and magnetic quantum numbers ($n = 0, 1, 2, \ldots$ and $m = -\infty, \ldots, +\infty$), respectively. Ω is known as the effective frequency of the system corresponding to

$$\Omega = \omega_0\left(1 + \left(\frac{\omega_c}{2\omega_0}\right)^2\right)^{\frac{1}{2}}. \tag{5}$$

Notice that, when the parameter $\omega_0 \to 0$, the energy levels of Equation (4) take the usual form of the Landau energy levels in cylindrical coordinates.

To obtain a more precise expression, especially when we consider the case of strong magnetic fields for the electron trapped in a quantum dot, we also take into account the electron spin of value $\frac{\hbar \hat{\sigma}}{2}$ and magnetic moment μ_B, where $\hat{\sigma}$ is the Pauli spin operator and $\mu_B = \frac{e\hbar}{2m^*}$. Here, the spin can be in two possible states, either $\uparrow$ or $\downarrow$, with respect to the applied external magnetic field B in the z-axis. Therefore, we include the Zeeman term in the Fock–Darwin energy levels in Equation (4). Consequently, the energy spectrum is given by

$$E_{n,m,\sigma} = \hbar\Omega(2n + |m| + 1) + m\frac{\hbar\omega_c}{2} - \mu_B \sigma B. \qquad (6)$$

The energy spectrum of Equation (6) is presented in Figure 1 for $\sigma = -1$ and $\sigma = 1$. It is interesting to note that, for high magnetic fields ($\omega_c/2\omega_0 \gg 1$), things simplify in Equation (6) and we get the following expression:

$$E_{n,m,\sigma} = \frac{\hbar\omega_c}{2}(n + 1/2 + |m| + m) - \mu_B \sigma B, \qquad (7)$$

where we observe that $|m| + m = 0$ for $m < 0$, therefore each Landau level labeled by n has infinite degeneracy.

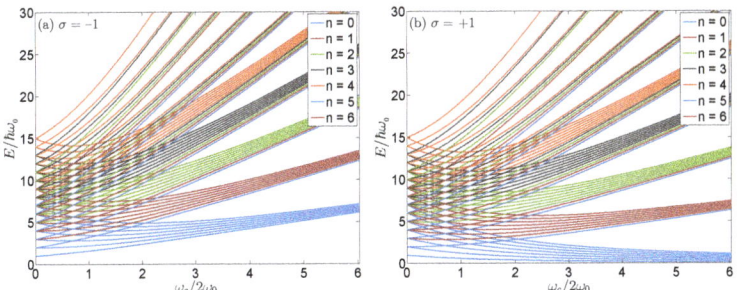

Figure 1. (a) Fock–Darwin energy spectrum with $\sigma = -1$ for the first six radial number $n = 0, 1, ..., 6$ and for each of them the azimuthal quantum number taking the values between $m = -6, -5, ..., 5, 6$. (b) Fock–Darwin energy spectrum with $\sigma = +1$ for the first six radial number $n = 0, 1, ..., 6$ and for each of them the azimuthal quantum number taking the values between $m = -6, -5, ..., 5, 6$. We clearly observe the confinement of the energy levels at high magnetic fields ($\omega_c/2\omega_0 \gg 1$).

In this paper, we consider a low-frequency coupling for the parabolic trap given by $\omega_0 \sim 2.637$ THz which in terms of energy units corresponds to a coupling of approximately 1.7 meV. The selection of this particular value is to compare the intensity of the trap with the typical energy of intra-band optical transitions of the quantum dots [47]. The order of this transition is approximately around ~ 1 meV for cylindrical GaAs quantum dots with effective mass given by $m^* \sim 0.067 \, m_e$ [47–49].

For the classical approach, we employ the framework of Refs. [50–53], and, in particular, classical thermodynamic quantities for the Fock–Darwin model with spin can be obtained analytically using the treatment of Kumar et al. [54]. For a working substance in thermal equilibrium at inverse temperature $\beta = 1/k_B T$, the partition function can be written as:

$$\mathcal{Z}_{dS} = \frac{1}{2}\mathrm{csch}\left(\frac{\hbar\beta\omega_+}{2}\right)\mathrm{csch}\left(\frac{\hbar\beta\omega_-}{2}\right)\cosh\left(\frac{\hbar\beta\omega_B}{2}\right), \qquad (8)$$

where the frequencies $\omega_\pm$ are:

$$\omega_\pm = \Omega \pm \frac{\omega_c}{2}. \qquad (9)$$

Therefore, entropy $(S(T,B))$, internal energy $(U(T,B))$ and magnetization $M(T,B)$ are simply given by

$$S(T,B) = k_B \ln \mathcal{Z}_{dS} + k_B T \left(\frac{\partial \ln \mathcal{Z}_{dS}}{\partial T}\right)_B, \qquad (10)$$

$$U(T,B) = k_B T^2 \left(\frac{\partial \ln \mathcal{Z}_{dS}}{\partial T}\right)_B, \qquad (11)$$

$$M(T,B) = k_B T \left(\frac{\partial \ln \mathcal{Z}_{dS}}{\partial B}\right). \qquad (12)$$

Equations (10)–(12) are presented in Figure 2 for a parabolic trap corresponding to an energy of 1.7 meV together with the scheme of the Otto cycle that we consider. A very interesting behavior is observed for the entropy as a function of the magnetic field in Figure 2a. For external magnetic fields ≤ 1 T, the entropy decreases as the external field increases, but for values higher than 1 T we see the opposite behavior. This can be explained by the energy levels becoming closer to each other as the magnetic field increases, moving towards degeneracy. This behavior in the energy levels causes the entropy growth as the magnetic field increases. In addition, the change in the behavior of the entropy is affected by temperature, finding that the change of slope as a function of external magnetic field moves away from the 1 T value to higher values as we move to higher temperature of the working substance. This can be appreciated in Figure 2a. At the same time, the magnetization shows a crossing in its behavior as a function of magnetic field, as we can see in Figure 2b, where previous to this crossing at lower temperatures higher values of magnetization are obtained. This fact becomes essential for the total work extracted. In the cycle that we propose, the work is directly related to the change in the magnetization of the system as a function of magnetic field and temperature. On the other hand, we can see that the internal energy monotonically decreases in terms of the magnetic field for all temperatures considered. The reason for this is that the internal energy only depends on the derivative of $\ln \mathcal{Z}_{dS}$ (see Equation (11)) with respect to temperature while the entropy has an additional term proportional to $\ln \mathcal{Z}_{dS}$ (see Equation (10)) and the magnetization on its derivative with respect to the external field (see Equation (12)).

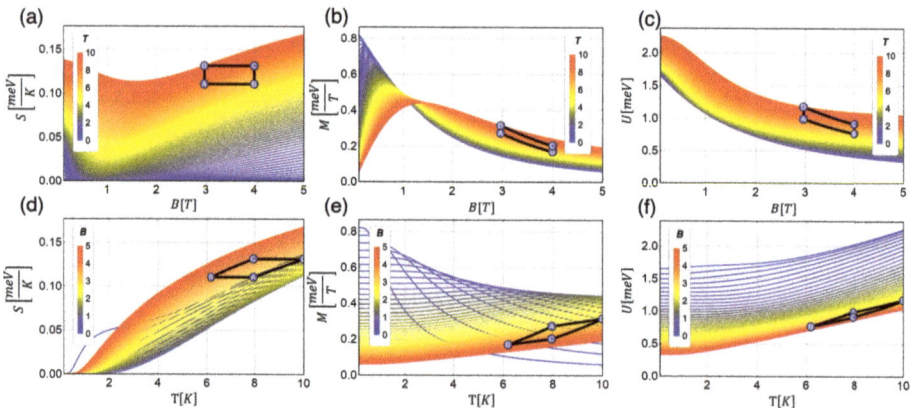

Figure 2. Classical thermodynamic quantities entropy (S), internal energy (U) and magnetization (M) as a function of: external magnetic field (B) (**a–c**); and temperature (T) (**d–f**). In (**a–c**), the colors blue to red represent temperatures from 0.1 K to 10 K, respectively. For (**d–f**), the colors blue to red represent lower to higher external magnetic field, from 0.1 T to 5 T. The value of the parabolic trap is approximately to 1.7 meV. Additionally, we show how the Otto cycle appears in terms of the thermodynamic quantities considered.

3. First Law of Thermodynamics and the Quantum and Classical Otto Cycle

The first law of thermodynamics in a quantum context has been discussed extensively in the literature. We follow the treatment in Refs. [50–52], which identifies the heat transferred and work performed during a thermodynamic process by means of the variation of the internal energy of the system.

First, consider a system described by a Hamiltonian, $\hat{\mathcal{H}}$, with discrete energy levels, $E_{n,m,\sigma}$. The internal energy of the system is simply the expectation value of the Hamiltonian $E = \langle \hat{\mathcal{H}} \rangle = \sum_n \sum_m \sum_\sigma p_{n,m,\sigma} E_{n,m,\sigma}$, where $p_{n,m,\sigma}$ are the corresponding occupation probabilities. The infinitesimal change of the internal energy can be written as

$$dE = \sum_n \sum_m \sum_\sigma (E_{n,m,\sigma} dp_{n,m,\sigma} + p_{n,m,\sigma} dE_{n,m,\sigma}), \tag{13}$$

where we can identify the infinitesimal work and heat as

$$dQ := \sum_n \sum_m \sum_\sigma E_{n,m,\sigma} dp_{n,m,\sigma}, \qquad dW := \sum_n \sum_m \sum_\sigma p_{n,m,\sigma} dE_{n,m,\sigma}. \tag{14}$$

Equation (13) is a formulation of the first law of thermodynamics for quantum working substances. Therefore, work is then related to a change in the eigenenergies $E_{n,m,\sigma}$, which is in agreement with the fact that work can only be carried out through a change in generalized coordinates. It is important to note that the expressions of Equation (14) is only a particular case of the definition of work and heat for a case of a density matrix operator that is diagonal on the energy eigenbasis [52]. A more complete definition of Equation (14) can be found in Refs. [29–33,46].

The quantum Otto cycle is composed of four strokes: two quantum isochoric processes and two quantum adiabatic processes. This cycle can be seen in Figure 3, replacing the value of S_l and S_h for $P_{n,m,\sigma}(T_l, B_h)$ and $P_{n,m,\sigma}(T_h, B_l)$ in the vertical axis, respectively. For the cases that we consider, the quantum Otto cycle proceeds as follows.

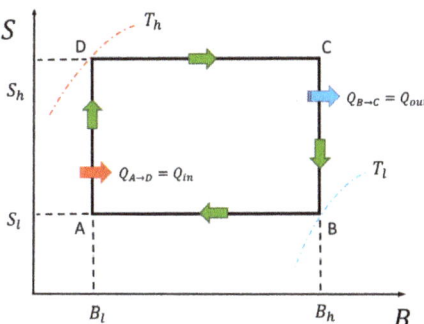

Figure 3. The magnetic Otto engine represented as an entropy (S) versus a magnetic field (B) diagram. The way to perform the cycle is in the form B → A → D → C → B.

1. Step B → A: Quantum adiabatic compression process. The systems, which is initialized in thermal equilibrium at temperature T_l, is isolated from the cold reservoir and the magnetic field is changed from B_h to B_l, with $B_h > B_l$. During this stage the populations remain constant, so $P_{n,m,\sigma}(T_l, B_h) = P^A_{n,m,\sigma}$. We remark that $P^A_{n,m,\sigma}$ does not yield a thermal state. No heat is exchanged during this process.

2. Step A → D: The system, at constant magnetic field B_l, is brought into contact with a hot thermal reservoir at temperature T_h until it reaches thermal equilibrium. The corresponding thermal

populations $P_{n,m,\sigma}(T_h, B_l)$ are given by the Boltzmann distribution with temperature T_h. No work is done during this stage.

The heat absorbed for the working substance is given by

$$Q_{in} = \sum_n \sum_m \sum_\sigma \int_A^D E_{n,m,\sigma} dP_{n,m,\sigma} = \sum_n \sum_m \sum_\sigma E_{n,m,\sigma}^l \left[P_{n,m,\sigma}(T_h, B_l) - P_{n,m,\sigma}^A \right], \qquad (15)$$

where $E_{n,m,\sigma}^l$ is the n-th eigenenergy of the system in the quantum isochoric heating process to an external magnetic field of value B_l.

3. Step D $\to$ C: Quantum adiabatic expansion process. The system is isolated from the hot reservoir, and the magnetic field is changed back from B_l to B_h. During this stage the populations remains constant, thus $P_{n,m,\sigma}(T_h, B_l) = P_{n,m,\sigma}^C$. Again, we remark that $P_{n,m,\sigma}^C$ is not a thermal state. No heat is exchanged during this process.

4. Step C $\to$ B: Quantum isochoric cooling process. The working substance at B_h is brought into contact with a cold thermal reservoir at temperature T_l. Therefore, the heat released is given by

$$Q_{out} = \sum_n \sum_m \sum_\sigma \int_C^B E_{n,m,\sigma} dP_{n,m,\sigma} = \sum_n \sum_m \sum_\sigma E_{n,m,\sigma}^h \left[P_{n,m,\sigma}(T_l, B_h) - P_{n,m,\sigma}^C \right], \qquad (16)$$

where $E_{n,m,\sigma}^h$ is the n-th eigenenergy of the system for an external magnetic field B_h.

The net work done in a single cycle can be obtained from $W = Q_{in} + Q_{out}$,

$$W = \sum_n \sum_m \sum_\sigma \left(E_{n,m,\sigma}^l - E_{n,m,\sigma}^h \right) \left(P_{n,m,\sigma}(T_h, B_l) - P_{n,m,\sigma}(T_l, B_h) \right), \qquad (17)$$

where we use the condition of constant populations along the quantum adiabatic strokes. Furthermore, the efficiency is given by

$$\eta = \frac{W}{Q_{in}}. \qquad (18)$$

The main difference between the classical and quantum Otto cycle is related to Points A and C in the cycle. In the classical case, the working substance can be at thermal equilibrium with a well-defined temperature at each point. On the other hand, for the quantum case, the working substance only reaches thermal equilibrium in the isochoric stages at Points B and D. After the adiabatic stages, the quantum system is in a diagonal state which is not a thermal state.

For the classical engine, the total work extracted by Equation (16) can be calculated by replacing $P_{n,m,\sigma}^A$ with $P(T_A, B_l)$ and $P_{n,m,\sigma}^C$ with $P(T_C, B_h)$, that is, it is obtained as a difference between the internal energy at adjacent points which can be calculated from the partition function

$$Q_{in} = U(T_h, B_l) - U(T_A, B_l); \qquad Q_{out} = U(T_l, B_h) - U(T_C, B_h), \qquad (19)$$

where T_A and T_C are determined by the condition imposed by the classical isentropic strokes. If we have the classical entropy, the intermediate temperatures T_A and T_C can be determined in two different forms:

- Finding the relation between the magnetic field and the temperature along an isentropic trajectory by solving the differential equation of first order given by

$$dS(B, T) = \left(\frac{\partial S}{\partial B} \right)_T dB + \left(\frac{\partial S}{\partial T} \right)_B dT = 0, \qquad (20)$$

which can be written as

$$\frac{dB}{dT} = -\frac{C_B}{T \left(\frac{\partial S}{\partial B} \right)_T}, \qquad (21)$$

where C_B is the specific heat at constant magnetic field.

- By matching two points within an isentropic trajectory

$$S(T_l, B_h) = S(T_A, B_l)$$
$$S(T_h, B_l) = S(T_C, B_h),$$
(22)

finding the magnetic field in terms of the temperature, throughout numerical calculation.

Therefore, from Equation (19) and $W = Q_{in} + Q_{out}$, the classical work is given by the difference of four internal energy in the form

$$W = U_D(T_h, B_l) - U_A(T_A, B_l) + U_B(T_l, B_h) - U_C(T_C, B_h),$$
(23)

It is important to mention that the cycle operation in the counter-clockwise form starting at Point A described in Figure 2 gives negative work extracted, thus, to define a thermal machine correctly, we start the cycle at Point B, and we go through it in a clockwise direction. This is due to the particular behavior of the entropy as a function of magnetic field and temperature in the chosen zone marked with A, B, C and D. Therefore, the cycle described in the next subsection is the form of B → A → D → C → B and is presented in Figure 3.

The maximum values considered in our calculations for the temperatures and external magnetic field were 10 K and 5 T. Therefore, for the quantum cycle calculation (i.e., Equation (17)), we used the quantum numbers $n = 0$ to $n = 10$ and $m = -33$ to $m = 33$ for Equation (6). The selection of this particular energy levels in this model is justified for the values of the thermal populations for the hot and cold temperatures of the reservoirs that we selected. Our numerical calculations indicated that the contributions of the other levels of energy can be neglected.

Finally, it is useful to express our results of total work extracted and efficiency in terms of the relation between the highest value (B_h) and the lowest value (B_l) of the external magnetic field over the sample. To do that, we used the definition of "magnetic length", which is given by

$$l_B = \sqrt{\frac{\hbar}{eB}},$$
(24)

allowing us to define the parameter

$$r = \frac{l_{B_l}}{l_{B_h}} = \sqrt{\frac{B_h}{B_l}},$$
(25)

which represents the analogy of the compression ratio for the classical case. It is important to remember that the Landau radius is inversely proportional to the magnitude of the magnetic field. Therefore, for a major (minor) magnitude of the field, the Landau radius is smaller (bigger), and the r is well defined.

4. Results and Discussions

4.1. Classical Magnetic Otto Cycle

The condition given by Equation (21) (or Equation (22)) for the classical cycle give us information about the behavior of the external magnetic field and the temperature in the adiabatic stroke. In Figure 4a, we can appreciate the level curves of the entropy function $S(T, B)$ and, Figure 4b shows some examples of isentropic strokes in a plot of $S(B)$ vs. B for different temperatures. That example shows three cases of constant low (red-black curve, $S = 0.05$), medium (yellow-black curve, $S = 0.10$) and high (white-black curve, $S = 0.13$) entropy. We observe in Figure 4a that there is a zone where the external field grows with the temperature of the sample and a zone where the opposite happens to maintain the entropy constant. At low working temperatures, the behavior changes near $B = 1$ T, while as the temperature increases, the slope change occurs at higher values of the magnetic field, approaching $B = 2$ T. Secondly, if we observe Figure 4b showing the case for $S = 0.13$ (white-black line), we have a restricted area for field values lower to 3 T if we work with a maximum temperature of 10 K. Therefore, the movement of the magnetic field is not arbitrary if we think in a thermodynamic magnetic Otto cycle with two temperature reservoirs fixed at some specific values, more specifically, the reservoir associated to the hot temperature in the cycle. In addition, Figure 4 is the solution of $S(T, B) = $ constant, obtained from the differential Equation (21) with different conditions (i.e., distinct values of the constant value of S). Therefore, Figure 4a depicts the entire family of solutions for the isentropic stroke of the engine of this particular system.

In our first example displayed in Figure 5, Point B has the value of the external field given by $B_h = 4$ T and a temperature of $T_B = 6.19$ K. The value of the temperature for Point D is fixed to $T_D = 10$ K. Therefore, the Carnot efficiency of the proposal cycle is given by

$$\eta_{Carnot} = 1 - \frac{T_B}{T_D} = 1 - \frac{6.19}{10} = 0.381 \qquad (26)$$

Figure 5e shows different values of total work extracted (W) varying the value of B_D from 4 T to 1.99 T. This variation in the external field is reflected in the movement of r in the form of $r = \sqrt{\frac{4}{B_1}}$. Therefore, r is in the range of $1 \leq r \leq 1.41$. The parabolic trap is fixed to the value of 1.7 meV and the effective mass in the value of $m^* = 0.067 m_e$. In particular, Figure 5a–c shows the exact paths for the magnetic cycle for the maximum point obtained when multiplying W (Figure 5e) and the efficiency (η, Figure 5f). That point corresponds to $r \sim 1.22$ (black point in Figure 5d–f) and constitutes the best configuration of the systems to obtain the best W with the better η through the cycle. In addition, W and η are presented in Figure 5e,f for the optimal value of r parameter mentioned before. We observe that W obtained for that point is in the order of ~0.038 meV with an efficiency of $\eta \sim 0.28$. We have corroborated the numeric result of total work extracted using the area enclosed by the cycle in Figure 5b of M versus B, as the work is $W = -\int MdB$ [50–52] when the parameter changed during the operation of the engine in the external field. On the other hand, to obtain the solid lines presented in Figure 5d–f, we needed to make different cycles configuration keeping the values of the isothermal fixed as can be appreciated in the *Supplementary Materials* (see the link after Section 5), made with the Mathematica software [55], where we show each shape that the cycle must have to generate a specific point of work. It is important to recall that we never reach the optimal value of $\eta = 0.381$, i.e., the Carnot efficiency.

Due to the change of behavior in the entropy as a function of the external field, we obtained very interesting results for W when we explored the zone close to $B = 1$ T. Before that point, the entropy decreases as function of the external field (B) and after that point entropy begins to increase. This fact can be used to explore the magnetic cycle in that zone finding an oscillatory behavior for W. In Figure 6, we show the cycle with operating temperatures $T_B = 2.69$ K and $T_D = 5.40$ K and external magnetic field moving between 2.995 T and 0.250 T and, consequently, the r parameter moving from 1 to 3.46. First, we observe a decreasing efficiency for $r > 1.75$ in Figure 6f with a maximum value of $\eta \sim 0.43$ for $r \sim 1.75$. Therefore, also for this configuration, the Carnot efficiency ($\eta_{Carnot} \sim 0.5$ for this case) cannot be reached. Comparing these results with those previously discussed (when we avoid this particular region), we can see in Figure 5f that the efficiency asymptotically approaches to the efficiency of Carnot if we increase the intensity of the external magnetic field of the starting point of the cycle (Point B).

In Figure 6b, we can understand the oscillations in W interpreting these results using the expression $W = -\int MdB$. In Figure 6a–c, Points A–D correspond to the black point displayed in Figure 6d–f where we see that the work is still greater than zero but close to a vanishing situation. The reason there is still positive work at this point under study is that the total area enclosed to the right of the crossing point is larger than the other to the left. The magnetization presented in Figure 6b in the zone around the range of external magnetic field explored for this case (from 2.995 to 0.250T) clearly reverses his behavior and presents a crossing point close to $B \sim 1.2$ T for different temperatures.

The area to the right of that point can be interpreted as a positive contribution to $\mathcal{W}$ while the left area contributes to negative work.

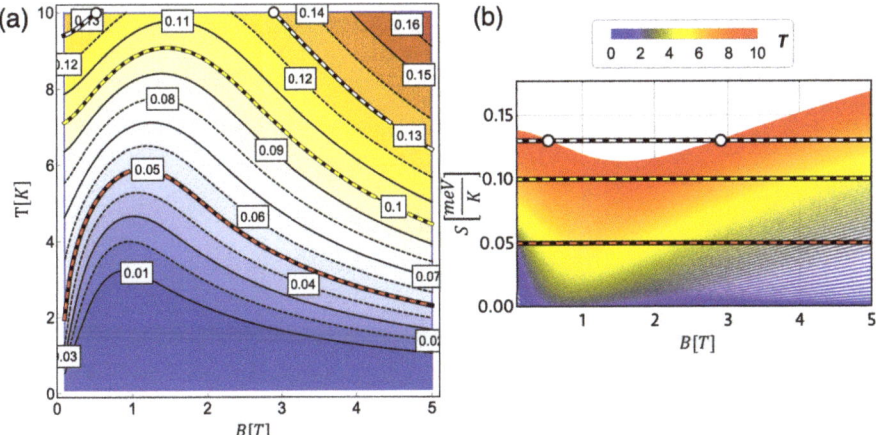

Figure 4. Solution of classical isentropic path. (**a**) The entropy as a function of magnetic field (horizontal axis) and temperature (vertical axis). The level curves (constant entropy values) highlight three different cases for S: first, red-black curve corresponding to $S = 0.05$; secondly, yellow-black curve, corresponding to $S = 0.10$ and finally, white-black curve for the case of $S = 0.13$. (**b**) The three constant values for the entropy ($S = 0.05, S = 0.10, S = 0.13$) in a graphic of entropy as a function of B for temperatures from 1 K (blue) up to 10 K (red). Due to the form of the entropy obtained for this system, the solution for $S = 0.13$ needs to work with temperatures higher than 10 K for an external magnetic field lower than 3 T (white dots in (**a**,**b**)). The value of the parabolic trap corresponds to 1.7 meV.

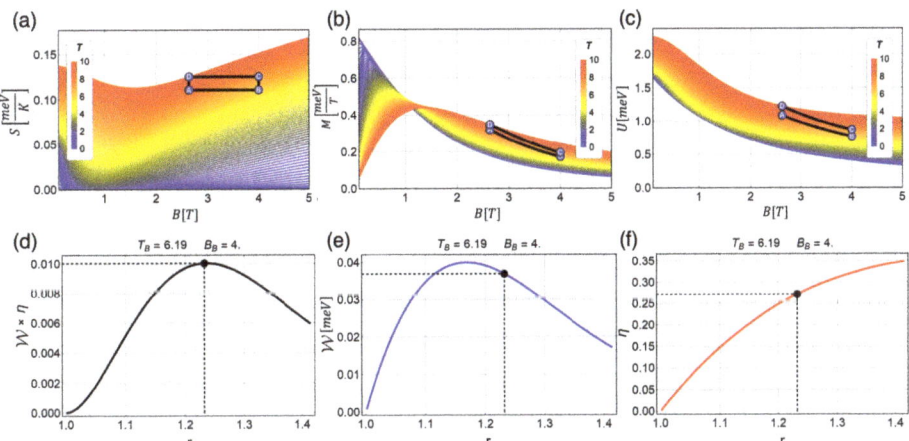

Figure 5. Proposed magnetic Otto cycle showing three different thermodynamic quantities, Entropy (S), Magnetization (M) and Internal Energy (U) ((**a**–**c**), respectively) as a function of the external magnetic field and different temperatures from $0.1K$ (blue) to $10K$ (red). (**d**) The total work extracted multiplied by efficiency ($\mathcal{W}\eta$); (**e**) the total work extracted ($\mathcal{W}$); and (**f**) the efficiency (η) for the classical cycle. The black points in (**d**–**f**) represent exactly the cycle B → A → D → C → B, presented in (**a**–**c**). The value of the parabolic trap corresponds to 1.7 meV. The fixed temperatures are $T_B = 6.19$ K and $T_D = 10$ K.

To explore if these oscillations in $\mathcal{W}$ are still obtained for higher temperature ranges, we plot in Figure 7 the work $\mathcal{W}$ for different values of T_D with $T_B = 2.69$ K fixed. We note that for higher temperatures than 7 K the oscillations found before disappear. It is only a reinforcement that the quantum effects of the working substance are only significant at low temperatures. On the other hand, as we expect, $\mathcal{W}$ grows as the difference between the temperature reservoir is larger, as shown in Figure 7a. However, for this case, the efficiency obtained is increasingly lower for increasingly larger temperature differences, as we can appreciate in Figure 7b.

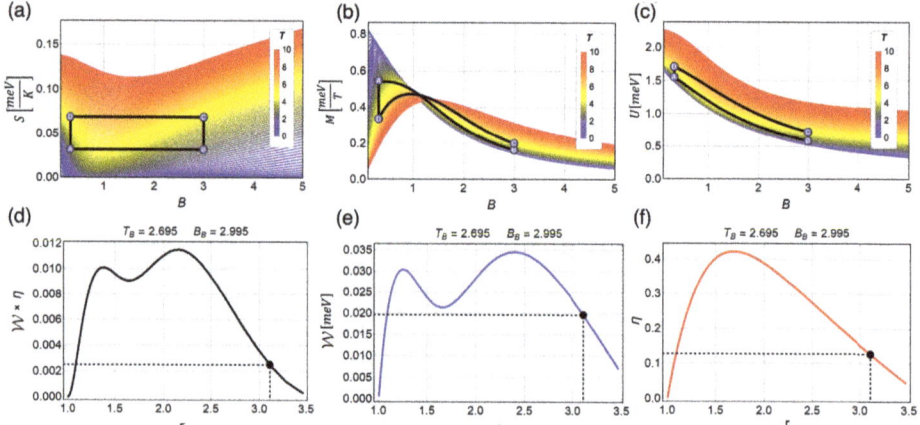

Figure 6. Proposed magnetic Otto cycle in three different thermodynamics quantities, Entropy, Magnetization and internal energy ((**a**–**c**), respectively) as a function of the external magnetic field and different temperatures from $0.1 K$ (blue) to $10 K$ (red). Total work extracted multiplied by efficiency ($W\eta$) (**d**) total work extracted (W) (**e**) and efficiency (η) (**d**) for the cycle. The black point in (**d**–**f**) represents the value of 0.02 meV of total work extracted and corresponds exactly to the cycle B → A → D → C → B, shown in (**a**–**c**). The value of the parabolic trap correspond to 1.7 meV. The fixed temperatures are $T_B = 2.69$ K and $T_D = 5.40$ K.

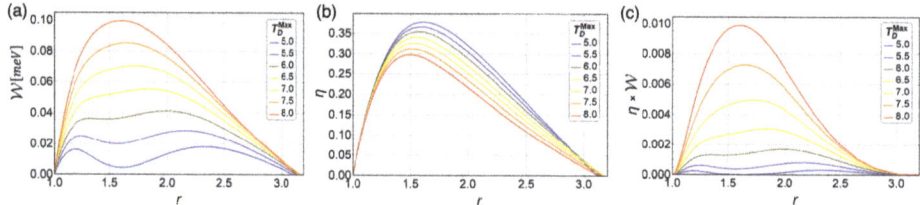

Figure 7. Work, efficiency and work multiply by efficiency (**a**–**c**) for different values of T_D for $T_B = 2.69$ fixed. The value of the parabolic trap corresponds to 1.7 meV.

4.2. Magnetic Quantum Otto Cycle

Next, we show the results of the evaluation of the quantum version of this magnetic Otto cycle for the same cases shown in Figures 5 and 6. In Figure 8a, we plot the classical work (blue line) and the quantum work (red line) for the same sets of parameters in Figure 5. First, we note that the classical and quantum work are equal up to the value of $r \sim 1.07$. This means, for values close to the starting external magnetic field to Point B, we do not notice a difference between the classic and quantum formulation of the Otto cycle. As shown in Figure 8a, we found a transition from positive work to negative work not reflected in the classic scenario close to $r \sim 1.26$.

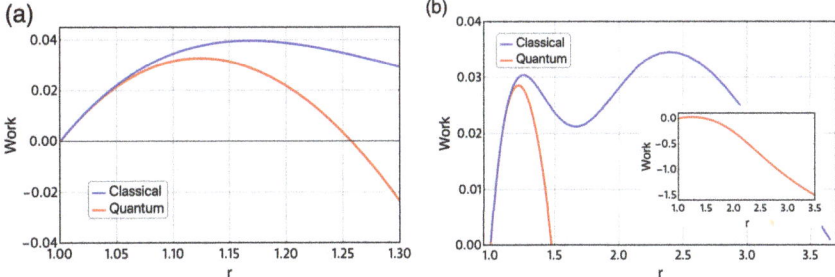

Figure 8. (a) Total work extracted for classical (blue line) and quantum version of Otto cycle (red line). The parameters for this case displayed are : $T_D = 10$ K, $T_B = 6.19$ K and $B_B = 4$ T as starting value of the external magnetic field. The value of B_D moves from 4 T to 1.99 T and this variation is reflected in the movement of r in the form of $r = \sqrt{\frac{4}{B_D}}$, same parameter as the results shown in Figure 5. (b) Total work extracted (W) presented in Figure 6e versus the values obtaining in the quantum version of the Otto cycle. The parameters for this figure are $T_D = 5.40$ K, $T_B = 2.69$ K and $B_B = 2.995$ T and B_D moves from 2.995 to 0.250 T. The parabolic trap is fixed to the value of 1.7 meV and the effective mass value of $m^* = 0.067 m_e$.

Additionally, we observe that the maximum positive value of the total work extracted for the quantum version of Otto cycle is reduced by approximately 0.01 meV compared to the classical counterpart. In particular, for the quantum version of this cycle, we did not found the oscillations in W presented in Figure 6e. Moreover, we found a transition from positive to negative work at some value of the r parameter. This is dramatically reflected in Figure 8b, where the absolute value of W is highly increased as compared with the classical approach.

In Figure 9, we present the work W per energy level and spin value for the most important values of our numerical calculations. We used the same parameter as in Figure 8b. We observed that the contribution given by $\sigma = 1$ are positive up to r close to $r \sim 1.6$ being the energy levels $E_{0,-1}$, $E_{0,-2}$ and $E_{0,-3}$ those that contribute with the most positive values. Contrarily, for the case of $\sigma = -1$, we found that all contributions per energy level are negative. Therefore, the small region of positive work found in Figure 8b can only be associated to the spin up ($\sigma = 1$) contributions.

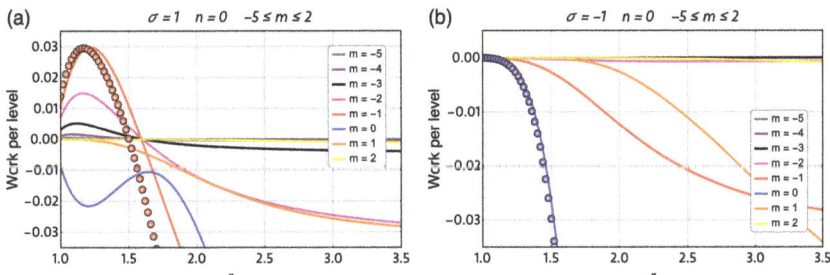

Figure 9. Total quantum work extracted (W) per energy level for the case of $\sigma = 1$ (a) and for the case of $\sigma = -1$ (b). The lines marked with circles correspond to the sum of all contributions of the energy level for each spin. The parameters used for this figure are the same as the one used in Figure 8b.

To explore other operation regions for the magnetic Otto cycle, we calculated the total work extracted and efficiency for the same $\Delta T = T_h - T_l$ in a broad range of temperatures and the same $\Delta B_{max} = 1.5$ T in different regions of the external magnetic field for the classical cycle and its quantum version. This is displayed in Figures 10 and 11 where the dotted lines represent the classical results and the solid lines the quantum results. The three regions of temperature selected for these two figures are 1–4 K (blue lines), 4–7 K (black lines) and 7–10 K (red lines). First, we treat the case of $B_B = 3.5$ T and B_D moving from 3.5 T to 2.0 T in Figure 10, where we note large differences between the classical and quantum results for W, as can be seen in Figure 10b. On the other hand, if we observed the region of $3.5 \leq B_D \leq 5.0$ T for a B_B fixed, as shown in Figure 11b. The work and efficiency for the region of 1–4 K and 4–7 K present similar behavior for the classical and quantum versions. Only the case of 7–10 K shows a larger difference between this two approaches. For the case of the efficiency, we note in Figure 10c a major difference between the classical results and quantum results compare with the presented in Figure 11c and this is consistent with the reported results for the work W.

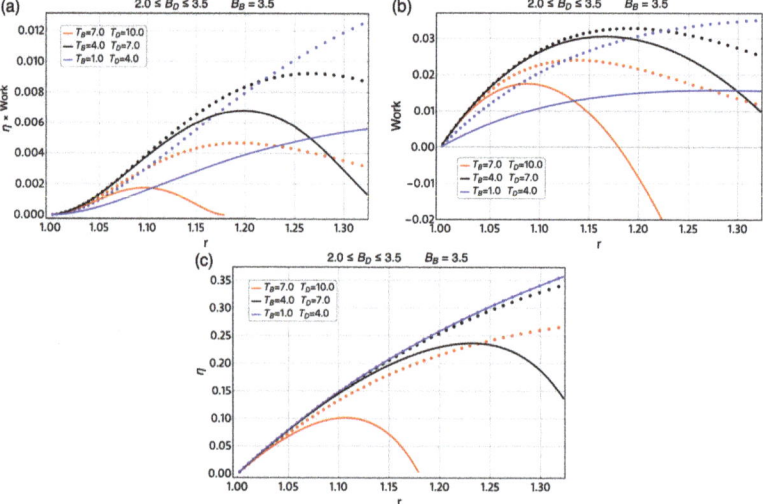

Figure 10. $\eta \times W$ (a); and total work extracted (b,c) efficiency for the case of $\Delta T = T_h - T_l = 3$ K for different regions of temperature parameter for classical approach (solid line) and quantum version of the magnetic Otto cycle (dotted line). For all graphics, we use the initial external magnetic field in the value of $B_B = 3.5$ T and the minimum value of the field, B_D moves between 3.5 T and 2.0 T. Therefore, the r parameter moves between $1 \leq r \leq 1.32$. The parabolic trap is fixed to the value of 1.7 meV and the effective mass value of $m^* = 0.067 m_e$.

Summarizing, our results show that it is the classical engine case with larger total work extracted and efficiency compared to the quantum formulation. This can be explained as follows.

The main difference between the classical and quantum version of Otto cycle lies in the fact that, in the classical formulation, the working substance can be in thermal equilibrium at each point in the cycle. In the quantum approach, the working substance is a single system that can only be in a thermal state after thermalizing with the reservoirs, which happens only in the isochoric strokes. After the adiabatic strokes, the working substance is in a diagonal state which is not a thermal state. In our case, the non-thermal points for the quantum case are Points C and A in Figure 3. The quantum work given by Equation (17) can be rewritten in the convenient form

$$\mathcal{W} = U_D(T_h, B_l) - \sum_{n,m,\sigma} E^l_{n,m,\sigma} P_{n,m,\sigma}(T_l, B_h) + U_D(T_l, B_h) - \sum_{n,m,\sigma} E^h_{n,m,\sigma} P_{n,m,\sigma}(T_h, B_l), \quad (27)$$

where, due to the thermal equilibrium of the two points (Points D and B in Figure 3), we can define the internal energy from equilibrium partition function. If we subtract the classical work given by Equation (23) from the quantum work written in the form of Equation (27), we obtain the following equation

$$\mathcal{W} - W = \left(\sum_{n,m,\sigma} E^l_{n,m,\sigma} P_{n,m,\sigma}(T_l, B_h) - U_A(T_A, B_l)\right) + \left(\sum_{n,m,\sigma} E^h_{n,m,\sigma} P_{n,m,\sigma}(T_h, B_l) - U_C(T_C, B_h)\right) \quad (28)$$

The first summation of Equation (28) is the average of the energy at low magnetic field with thermal probabilities that satisfies the adiabatic condition

$$S = -k_B \sum_{n,m,\sigma} P_{n,m,\sigma}(T_l, B_h) \ln\left(P_{n,m,\sigma}(T_l, B_h)\right), \qquad (29)$$

i.e., the entropy at Point A. On the other hand, $U_A(T_A, B_l)$ is the average value of the energy at low external field and in thermal equilibrium, with the same value of entropy presented in Equation (29). Therefore, $U_A(T_A, B_l)$ is the absolute minimum according to thermodynamic [53]. The same argument can be made at Point C, thus the difference of classical work minus quantum work always satisfies the following condition

$$\mathcal{W} - W \geq 0 \qquad (30)$$

This result applies to any system in which the occupation probabilities of the energy levels at any magnetic field are replaced with any form, provided that they satisfy the adiabatic condition. This is because the value at equilibrium of any internal parameter (without constrains) of the system, makes the internal energy to be a minimum for a given value of the total Entropy [53].

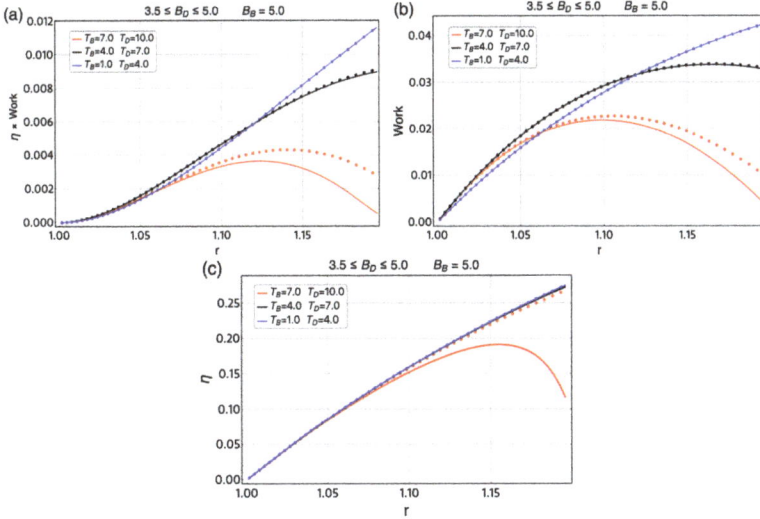

Figure 11. $\eta \times W$ (a); and total work extracted (b,c) efficiency for the case of $\Delta T = T_h - T_l = 3\ K$ for different regions of temperature parameter. For all cases, we use the initial external magnetic field at the value of $B_B = 5.0$ T and the minimum value of the field, B_D moves between 5.0 T and 3.5 T. Therefore, the r parameter moves between $1 \leq r \leq 1.19$. The parabolic trap is fixed to the value of 1.7 meV and the effective mass value of $m^* = 0.067 m_e$.

5. Conclusions

In this work, we explored the classical and quantum approach for a magnetic Otto cycle for an ensemble of non interacting electrons with intrinsic spin where each one is trapped inside a semiconductor quantum dot modeled by a parabolic potential. We analyzed all relevant thermodynamics quantities, and found that the entropy changes it behavior in terms of the external magnetic field at the point where the energy spectrum tends towards degeneracy; this behavior was present at all temperatures considered. This behavior determined the range of parameters such as temperature and external magnetic field that would lead to the operation of the Otto cycle extracting positive total work. In the classical approach, we found oscillations in the total work extracted that are not present in the quantum approach. This happened near the zone of slope change in the behavior of the entropy in terms of the magnetic field. Interestingly, we found that, in the classical approach, the engine yielded a much higher performance in terms of total work extracted and efficiency than in the quantum approach. This is because, in the classical approach, the working substance can be in thermal equilibrium at each point of the cycle, whereas, in the quantum approach, the working substance can only thermalize in the isochoric strokes. Because of the principle of minimum energy, the system is allowed to extract more energy when the adiabatic strokes can lead to states that are in thermal equilibrium, which is only possible in the classical case.

These results are reasonable, since, in our quantum approach, the working substance remains in a diagonal state and does not use quantum resources such as quantum coherence, which in some cases can lead to enhanced performance.

Supplementary Materials: The following are available online at http://www.mdpi.com/1099-4300/21/5/512/s1, Video S1: "Work evolution for high field/temperature zones", Video S2: "Oscillatory behaviour of classical work extracted I", Video S3: "Oscillatory behaviour of classical work extracted II". Video S1 shows the behaviour of work and efficiency in the high field/temperature zones for the proposed machine. Video S2 and S3 shows the oscillatory nature of the extracted work for the classical version of the Otto cycle due to the entropy behavior.

Author Contributions: F.J.P., G.A. and P.V. conceived the idea and formulated the theory. O.N., D.Z and A.G. built the computer program and edited figures. A.S.N. and P.A.O. contributed to discussions during the entire work and editing the manuscript. F.J.P. wrote the paper. All authors have read and approved the final manuscript.

Funding: This research was funding by Financiamiento Basal para Centros Científicos y Tecnológicos de Excelencia, under Project No. FB 0807 (Chile).

Acknowledgments: F.J.P. acknowledges the financial support of FONDECYT-postdoctoral 3170010, and D.Z. acknowledges USM-DGIIP. P.V. and G.A.B. acknowledges support from Financiamiento Basal para Centros Científicos y Tecnológicos de Excelencia, under Project No. FB 0807 (Chile). The authors acknowledge DTI-USM for the use of "Mathematica Online Unlimited Site" at the Universidad Técnica Federico Santa María.

Conflicts of Interest: The authors declare no conflict of interest.

References

1. Scovil, H.E.D.; Schulz-DuBois, D.O. Three-Level masers as a heat engines. *Phys. Rev. Lett.* **1959**, *2*, 262–263. [CrossRef]
2. Feldmann, T.; Geva, E.; Kosloff, R.; Salamon, P. Heat engines in finite time governed by master equations. *Am. J. Phys.* **1996**, *64*, 485. [CrossRef]
3. Feldmann, T.; Kosloff, R. Characteristics of the limit cycle of a reciprocating quantum heat engine. *Phys. Rev. E* **2004**, *70*, 046110. [CrossRef]
4. Rezek, Y.; Kosloff, R. Irreversible performance of a quantum harmonic heat engine. *New J. Phys.* **2006**, *8*, 83. [CrossRef]
5. Henrich, M.J.; Rempp, F.; Mahler, G. Quantum thermodynamic Otto machines: A spin-system approach. *Eur. Phys. J. Spec. Top.* **2007**, *151*, 157. [CrossRef]
6. Quan, H.T.; Liu, Y.-X.; Sun, C.P.; Nori, F. Quantum thermodynamic cycles and quantum heat engines. *Phys. Rev. E* **2007**, *76*, 031105. [CrossRef]
7. He, J.; He, X.; Tang, W. The performance characteristics of an irreversible quantum Otto harmonic refrigeration cycle. *Sci. China Ser. G Phy. Mech. Astron.* **2009**, *52*, 1317. [CrossRef]
8. Liu, S.; Ou, C. Maximum Power Output of Quantum Heat Engine with Energy Bath. *Entropy* **2016**, *18*, 205. [CrossRef]
9. Scully, M.O.; Zubairy, M.S.; Agarwal, G.S.; Walther, H. Extracting work from a single heath bath via vanishing quantum coherence. *Science* **2003**, *299*, 862–864. [CrossRef]
10. Scully, M.O.; Zubairy, M.S.; Dorfmann, K.E.; Kim, M.B.; Svidzinsky, A. Quantum heat engine power can be increased by noise-induced coherence. *Proc. Natl. Acad. Sci. USA* **2011**, *108*, 15097–15100. [CrossRef]
11. Bender, C.M.; Brody, D.C.; Meister, B.K. Quantum mechanical Carnot engine. *J. Phys. A Math. Gen.* **2000**, *33*, 4427. [CrossRef]
12. Bender, C.M.; Brody, D.C.; Meister, B.K. Entropy and temperature of quantum Carnot engine. *Proc. R. Soc. Lond. A* **2002**, *458*, 1519. [CrossRef]

13. Wang, J.H.; Wu, Z.Q.; He, J. Quantum Otto engine of a two-level atom with single-mode fields. *Phys. Rev. E* **2012**, *85*, 041148. [CrossRef]
14. Huang, X.L.; Xu, H.; Niu, X.Y.; Fu, Y.D. A special entangled quantum heat engine based on the two-qubit Heisenberg XX model. *Phys. Scr.* **2013**, *88*, 065008. [CrossRef]
15. Muñoz, E.; Peña, F.J. Quantum heat engine in the relativistic limit: The case of Dirac particle. *Phys. Rev. E* **2012**, *86*, 061108. [CrossRef] [PubMed]
16. Quan, H.T. Quantum thermodynamic cycles and quantum heat engines (II). *Phys. Rev. E* **2009**, *79*, 041129. [CrossRef]
17. Zheng, Y.; Polleti, D. Work and efficiency of quantum Otto cycles in power-law trapping potentials. *Phys. Rev. E* **2014**, *90*, 012145. [CrossRef]
18. Cui, Y.Y.; Chem, X.; Muga, J.G. Transient Particle Energies in Shortcuts to Adiabatic Expansions of Harmonic Traps. *J. Phys. Chem. A* **2016**, *120*, 2962. [CrossRef] [PubMed]
19. Beau, M.; Jaramillo, J.; del Campo, A. Scaling-up Quantum Heat Engines Efficiently via Shortcuts to Adiabaticity. *Entropy* **2016**, *18*, 168. [CrossRef]
20. Deng, J.; Wang, Q.; Liu, Z.; Hänggi, P.; Gong, J. Boosting work characteristics and overall heat-engine performance via shortcuts to adibaticity: Quantum and classical systems. *Phys. Rev. E* **2013**, *88*, 062122. [CrossRef] [PubMed]
21. Wang, J.; He, J.; He, X. Performance analysis of a two-state quantum heat engine working with a single-mode radiation field in a cavity. *Phys. Rev. E* **2011**, *84*, 041127. [CrossRef]
22. Abe, S. Maximum-power quantum-mechanical Carnot engine. *Phys. Rev. E* **2011**, *83*, 041117. [CrossRef]
23. Wang, J.H.; He, J.Z. Optimization on a three-level heat engine working with two noninteracting fermions in a one-dimensional box trap. *J. Appl. Phys.* **2012**, *111*, 043505. [CrossRef]
24. Wang, R.; Wang, J.; He, J.; Ma, Y. Performance of a multilevel quantum heat engine of an ideal N-particle Fermi system. *Phys. Rev. E* **2012**, *86*, 021133. [CrossRef]
25. Jaramillo, J.; Beau, M.; del Campo, A. Quantum supremacy of many-particle thermal machines. *New J. Phys.* **2016**, *18*, 075019. [CrossRef]
26. del Campo, A.; Goold, J.; Paternostro, M. More bang for your buck: Super-adiabatic quantum engines. *Sci. Rep.* **2017**, *4*, 14391.
27. Huang, X.L.; Niu, X.Y.; Xiu, X.M.; Yi, X.X. Quantum Stirling heat engine and refrigerator with single and coupled spin systems. *Eur. Phys. J. D* **2014**, *68*, 32. [CrossRef]
28. Su, S.H.; Luo, X.Q.; Chen, J.C.; Sun, C.P. Angle-dependent quantum Otto heat engine based on coherent dipole-dipole coupling. *EPL* **2016**, *115*, 30002. [CrossRef]
29. Kosloff, R.; Rezek, Y. The Quantum Harmonic Otto Cycle. *Entropy* **2017**, *19*, 136. [CrossRef]
30. Klatzow, J.; Becker, J.N.; Ledingham, P.M.; Weinzetl, C.; Kaczmarek, K.T.; Saunders D.J.; Nunn J.; Walmsley, I.A.; Uzdin, R.M.; Poem, E. Experimental demonstration of quantum effects in the operation of microscopic heat engines. *arXiv* **2017**, arXiv:1710.08716v2.
31. Peterson J.P.S.; Batalhão T.B.; Herrera, M.; Souza A.M.; Sarthour R.S.; Oliveira I. S.; Serra, R.M. Experimental characterization of a spin quantum heat engine. *arXiv* **2018**, arXiv:1803.06021v1.
32. von Lindenfels, D.; Gräb, O.; Schmiegelow, C.T.; Kaushal, V.; Schulz, J.; Schmidt-Kaler, F.; Poschinger, U.G. A spin heat engine coupled to a harmonic-oscillator flywheel. *arXiv* **2018**, arXiv:1808.02390v1.
33. Van Horne, N.; Yum, D.; Dutta, T.; Hänggi, P.; Gong, J.; Poletti, D.; Mukherjee, M. Single atom energy-conversion device with a quantum load. *arXiv* **2018**, arXiv:1808.02390v1.
34. Roßnagel, J.; Dawkins, T.K.; Tolazzi, N.K.; Abah, O.; Lutz, E.; Kaler-Schmidt, F.; Singer, K.A. Single-atom heat engine. *Science* **2016**, *352*, 325. [CrossRef]
35. Dong, C.D.; Lefkidis, G.; Hübner, W. Quantum Isobaric Process in Ni_2. *J. Supercond. Nov. Magn.* **2013**, *26*, 1589–1594. [CrossRef]
36. Dong, C.D.; Lefkidis, G.; Hübner, W. Magnetic quantum diesel in Ni_2. *Phys. Rev. B* **2013**, *88*, 214421. [CrossRef]
37. Hübner, W.; Lefkidis, G; Dong, C.D.; Chaudhuri, D. Spin-dependent Otto quantum heat engine based on a molecular substance. *Phys. Rev. B* **2014**, *90*, 024401. [CrossRef]
38. Mehta, V.; Johal, R.S. Quantum Otto engine with exchange coupling in the presence of level degeneracy. *Phys. Rev. E* **2017**, *96*, 032110. [CrossRef]

39. Azimi, M.; Chorotorlisvili, L.; Mishra, S.K.; Vekua, T.; Hübner, W.; Berakdar, J. Quantum Otto heat engine based on a multiferroic chain working substance. *New J. Phys.* **2014**, *16*, 063018. [CrossRef]
40. Chotorlishvili, L.; Azimi, M.; Stagraczyński, S.; Toklikishvili, Z.; Schüler, M.; Berakdar, J. Superadiabatic quantum heat engine with a multiferroic working medium. *Phys. Rev. E* **2016**, *94*, 032116. [CrossRef]
41. Muñoz, E.; Peña, F.J. Magnetically driven quantum heat engine. *Phys. Rev. E* **2014**, *89*, 052107. [CrossRef]
42. Alecce, A.; Galve, F.; Gullo, N.L.; Dell'Anna, L.; Plastina, F.; Zambrini, R. Quantum Otto cycle with inner friction: Finite-time and disorder effects. *New J. Phys.* **2015**, *17*, 075007. [CrossRef]
43. Brandner, K.; Bauer, M.; Seifert, U. Universal Coherence-Induced Power Losses of Quantum Heat Engines in Linear Response. *Phys. Rev. Lett.* **2017**, *119*, 170602. [CrossRef]
44. Feldmann, T.; Koslof, R. Performance of discrete heat engines and heat pumps in finite time. *Phys. Rev. E* **2000**, *61*, 4774. [CrossRef]
45. Feldmann, T.; Koslof, R. Transitions between refrigeration regions in extremely short quantum cycles. *Phys. Rev. E* **2016**, *93*, 052150. [CrossRef] [PubMed]
46. Pekola, J.P.; Karimi, B; Thomas, G.; Averin, D.V. Supremacy of incoherent sudden cycles cycles. *arXiv* **2018**, arXiv:1812.10933v1.
47. Jacak, L.; Hawrylak, P.; Wójs, A. *Quantum Dots*; Springer: New York, NY, USA, 1998.
48. Muñoz, E.; Barticevic, Z.; Pacheco, M. Electronic spectrum of a two-dimensional quantum dot array in the presence of electric and magnetic fields in the Hall configuration. *Phys. Rev. B* **2005** *71*, 165301. [CrossRef]
49. Mani, R.G.; Smet, J.H.; von Klitzing, K.; Narayanamurti, V.; Johnson, W.B.; Umansky, V. Zero-resistance states induced by electromagnetic-wave excitation in GaAs/AlGaAs heterostructures. *Nature* **2002**, *420*, 646–650. [CrossRef]
50. Quan, H.T.; Zhang, P.; Sun, C.P. Quantum heat engine with multilevel quantum systems. *Phys. Rev. E* **2005**, *72*, 056110. [CrossRef]
51. Peña, F.J.; Muñoz, E. Magnetostrain-driven quantum heat engine on a graphene flake. *Phys. Rev. E* **2015**, *91*, 052152. [CrossRef]
52. Muñoz, E.; Peña, F.J.; González, A. Magnetically-Driven Quantum Heat Engines: The Quasi-Static Limit of Their Efficiency. *Entropy* **2016**, *18*, 173. [CrossRef]
53. Callen, H.B. *Thermodynamics and an Introduction to Thermostatistics*; John Wiley & Sons: New York, NY, USA, 1985.
54. Kumar, J.; Sreeram P.A.; Dattagupta, S. Low-temperature thermodynamics in the context of dissipative diamagnetism. *Phys. Rev. E* **2009**, *79*, 021130. [CrossRef]
55. Wolfram Research, Inc. *Mathematica*, Version 11.3, Wolfram Research, Inc.: Champaign, IL, USA, 2018.

© 2019 by the authors. Licensee MDPI, Basel, Switzerland. This article is an open access article distributed under the terms and conditions of the Creative Commons Attribution (CC BY) license (http://creativecommons.org/licenses/by/4.0/).

Article

Non-Thermal Quantum Engine in Transmon Qubits

Cleverson Cherubim [1,*], Frederico Brito [1] and Sebastian Deffner [2]

[1] Instituto de Física de São Carlos, Universidade de São Paulo, C.P. 369, 13560-970 São Carlos, SP, Brazil; fbb@ifsc.usp.br
[2] Department of Physics, University of Maryland Baltimore County, Baltimore, MD 21250, USA; deffner@umbc.edu
* Correspondence: cleverson.cherubim@usp.br

Received: 8 May 2019; Accepted: 27 May 2019; Published: 29 May 2019

Abstract: The design and implementation of quantum technologies necessitates the understanding of thermodynamic processes in the quantum domain. In stark contrast to macroscopic thermodynamics, at the quantum scale processes generically operate far from equilibrium and are governed by fluctuations. Thus, experimental insight and empirical findings are indispensable in developing a comprehensive framework. To this end, we theoretically propose an experimentally realistic quantum engine that uses transmon qubits as working substance. We solve the dynamics analytically and calculate its efficiency.

Keywords: quantum heat engines; quantum thermodynamics; nonequilibrium systems

1. Introduction

Recent advances in nano and quantum technology will necessitate the development of a comprehensive framework for *quantum thermodynamics* [1]. In particular, it will be crucial to investigate whether and how the laws of thermodynamics apply to small systems, whose dynamics are governed by fluctuations and which generically operate far from thermal equilibrium. In addition, it has already been recognized that at the nanoscale many standard assumptions of classical statistical mechanics and thermodynamics are no longer justified and even in equilibrium quantum subsystems are generically not well-described by a Maxwell-Boltzmann distribution, or rather a Gibbs state [2]. Thus, the formulation of the statements of quantum thermodynamics have to be carefully re-formulated to account for potential quantum effects in, for instance, the efficiency of heat engines [3–6].

In good old thermodynamic tradition, however, this conceptual work needs to be guided by experimental insight and empirical findings. To this end, a cornerstone of quantum thermodynamics has been the description of the working principles of quantum heat engines [7–17].

However, to date it is not unambiguously clear whether quantum features can always be exploited to outperform classical engines, since to describe the thermodynamics of non-thermal states one needs to consider different perspectives—different than the one established for equilibrium thermodynamics. For instance, it has been shown that the Carnot efficiency cannot be beaten [4,18] if one accounts for the energy necessary to maintain the non-thermal stationary state [19–22]. However, it has also been argued that Carnot's limit can be overcome, if one carefully separates the "heat" absorbed from the environment into two different types of energy exchange [23,24]: one is associated with a variation in *passive energy* [25,26] which would be the part responsible for changes in entropy, and the other type is a variation in *ergotropy*, a work-like energy that could be extracted by means of a suitable unitary transformation. On the other hand, it has been shown [27] that a complete thermodynamic description in terms of *ergotropy* is also not always well suited. Having several perspectives to explain the same phenomenon is a clear indication of the subtleties and challenges faced by quantum thermodynamics, and which can only be settled by the execution of purposefully designed experiments. Therefore, theoretical proposals for feasible and relevant experiments appear instrumental.

In this work we propose an experiment to implement a thermodynamic engine with a transmon qubit as the working substance (WS), which interacts with a non-thermal environment composed by two subsystems, an externally excited cavity (a superconducting transmission line) and a classical heat bath [28] with temperature T. The WS undergoes a non-conventional cycle (different from Otto, Carnot, etc.) [29] through a succession of non-thermal stationary states obtained by slowly varying its bare energy gap (frequency) and the amplitude of the pumping field applied to the cavity. We calculate the efficiency of this engine for a range of experimentally accessible parameters [28,30–32], obtaining a maximum value of 47%, which is comparable with values from the current literature.

2. System Description

We consider a multipartite system, comprised of a transmon qubit of tunable frequency ω_T, which interacts with a transmission line (cavity) of natural frequency ω_{CPW} with coupling strength g. The cavity is pumped by an external field of amplitude E_d and single frequency ω (see Figure 1). Both systems are in contact with a classical heat bath at temperature T. Such a set-up is experimentally realistic and several implementations have already been reported in different contexts [28,33]. Here and in the following, the transmon is used as a working substance (WS) and the (non-standard) "bath" is represented by the net effect of the other two systems: the cavity and the cryogenic environment (classical bath). There are two subtleties that must be noted here: (i) the bath "seen" by the qubit does not only consist of a classical reservoir at some fixed temperature, but it has an additional component, namely the pumped cavity. By changing the pumping, several cavity states can be realized. Such a feature gives the possibility of making this composed bath *non-thermal* on demand. In addition, (ii), the proposed engine is devised as containing only one bath (cavity + environment), which does not pose any problems considering that it is an out-of-equilibrium bath.

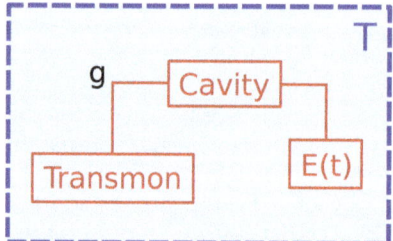

Figure 1. Sketch of the quantum engine with a transmon qubit as working substance interacting with an externally pumped (E(t)) transmission line (cavity). Both systems are embedded in the same cryogenic environment, which plays the role of a standard thermal bath of temperature T. Such a setup gives a dynamics of a working substance in the presence of a controllable *non-thermal* environment.

We start our analysis from the Hamiltonian describing a tunable qubit interacting with a single mode pumped cavity through a Jaynes-Cummings interaction

$$H(t) = \frac{\hbar\omega_T}{2}\sigma_z + \hbar\omega_{CPW}a^\dagger a + g\sigma_x(a + a^\dagger) \\ + E_d\left(ae^{i\omega t} + a^\dagger e^{-i\omega t}\right), \quad (1)$$

where σ_x and σ_z are the Pauli matrices, $a^\dagger$ and a are the canonical bosonic creation and annihilation operators associated with the cavity excitations, g is the qubit-cavity coupling strength. The last term represents a monochromatic pumping of amplitude E_d and frequency ω applied to the cavity.

The experimental characterization of the qubit-cavity dissipative dynamics emerging from their interaction with the same thermal bath shows that the system's steady state is determined by the master equation [28]

$$\dot{\rho}(t) = -\frac{i}{\hbar}[H_{\text{RWA}}, \rho] + K_{\text{CPW}}^- \mathcal{D}[a]\rho \\ + K_{\text{CPW}}^+ \mathcal{D}[a^\dagger]\rho + \Gamma^- \mathcal{D}[\sigma^-]\rho + \Gamma^+ \mathcal{D}[\sigma^+]\rho, \quad (2)$$

with K_{CPW}^- (K_{CPW}^+) being the cavity decay (excitation) rate, Γ^- (Γ^+) the qubit relaxation (excitation) rate and $\mathcal{D}[A]\rho = A\rho A^\dagger - 1/2(A^\dagger A \rho + \rho A^\dagger A)$. Please note that these rates satisfy detailed balance for the same bath of temperature T, $K_{\text{CPW}}^+/K_{\text{CPW}}^- = \exp(-\hbar\omega_{\text{CPW}}/k_B T)$ and $\Gamma^+/\Gamma^- = \exp(-\hbar\omega_T/k_B T)$. The Hamiltonian part

$$H_{\text{RWA}} = \frac{\hbar}{2}(\omega_T - \omega)\sigma_z + \hbar(\omega_{\text{CPW}} - \omega)a^\dagger a \\ + g(\sigma_+ a + \sigma_- a^\dagger) + E_d(a + a^\dagger), \quad (3)$$

is the system Hamiltonian in the rotating wave approximation (RWA) [34], with σ_+ (σ_-) being the spin ladder operators.

Since we are interested in the observed dynamics of the WS, it is necessary to find the qubit's reduced density matrix $\rho_T(t) \equiv \text{tr}_a\{\rho(t)\}$, where $\text{tr}_a\{\cdot\}$ represents the partial trace on the cavity's degrees of freedom. The system state is in a qubit-cavity product state, i.e., $\rho(t) \approx \rho_T(t) \otimes \rho_C(t)$, which emerges in the effective qubit-cavity weak coupling regime due to decoherence into the global environment. In addition, the cavity's stationary state $\rho_C(t)$ is assumed to be mainly determined by the external pumping, which can be easily found for situations of strong pumping and/or weak coupling strength g. This closely resembles a situation, in which the cavity acts as a work source of effectively infinite inertia [35]. Thus, changing the state of the qubit does not affect the state of the cavity, but it is still susceptible to the applied field and the cryogenic bath, and we have

$$\langle a \rangle = \langle a^\dagger \rangle^* = \frac{E_d}{\hbar[i\kappa_{\text{CPW}}/2 - (\omega_{\text{CPW}} - \omega)]}, \quad (4)$$

where we defined $K_{\text{CPW}}^- = \kappa_{\text{CPW}}$. Hence, the reduced master equation (2) can be written as

$$\dot{\rho}_T(t) = -\frac{i}{\hbar}[\tilde{H}_{T,\text{RWA}}, \rho_T] + \Gamma^- \mathcal{D}[\sigma^-]\rho_T + \Gamma^+ \mathcal{D}[\sigma^+]\rho_T, \quad (5)$$

with

$$\tilde{H}_{T,\text{RWA}} = \frac{\hbar}{2}(\omega_T - \omega)\sigma_z + g\left[\langle a \rangle \sigma_+ + \langle a^\dagger \rangle \sigma_-\right]. \quad (6)$$

Please note that the effective qubit Hamiltonian carries information about the interaction with the cavity through $\langle a \rangle$ and $\langle a^\dagger \rangle$, which are dependent on the cavity state.

3. Non-Equilibrium Thermodynamics

3.1. Non-Thermal Equilibrium States

The only processes that are fully describable by means of conventional thermodynamics are infinitely slow successions of equilibrium states. For the operating principles of heat engines, the second law states that the maximum attainable efficiency of a thermal engine operating between two heat baths is limited by Carnot's efficiency.

An extension of this standard description is considering infinitely slow successions along non-Gibbsian, but stationary states [4,18–20,36]. In the present case, namely a heat engine with transmon qubit as working substance, non-Gibbsianity is induced by the external excitation applied as a driving field to the cavity. We will see in the following, however, that identifying the thermodynamic work is subtle – and that the energy exchange can exhibit heat-like character, which is crucial when computing the entropy variation during the engine operation.

The stationary state can be found by solving the master equation Equation (5), and is written as

$$\rho_T^{ss} = \begin{pmatrix} \rho_T^{ee} & \rho_T^{eg} \\ \rho_T^{ge} & \rho_T^{gg} \end{pmatrix} \quad (7)$$

where the matrix elements can be computed explicitly and are summarized in Appendix A.

We observe that for the case of effective qubit-cavity ultra-weak coupling, i.e., $\hbar\omega_T \gg gE_d/|i\hbar\kappa_{CPW}/2 - \hbar(\omega_{CPW} - \omega)|$, as expected, the obtained non-thermal state asymptotically approaches thermal equilibrium, namely $|\rho_T^{eg}| = |\rho_T^{ge}| \approx 0$ and $\rho_T^{ee}/\rho_T^{gg} \approx \exp(-\beta\hbar\omega_T)$. In addition, as also expected, in the high temperature limit $\hbar\omega_T/kT \ll 1$ the qubit stationary state becomes the thermal, maximally mixed state, given that the cavity is not strongly pumped.

3.2. The Cycle

In equilibrium thermodynamics cycles are constructed by following a closed path on a surface obtained by the equation of state [29], which characterizes possible equilibrium states for a given set of macroscopic variables. This procedure can be generalized in the context of steady state thermodynamics, where an equation of state is also constructed.

For the present purposes, we use the steady state (7) to devise a cycle for our heat engine. The equation of state in our case is represented by the stationary state's von Neumann entropy $S(\omega_T, E_d) = -\text{tr}\{\rho_T^{ss} \ln \rho_T^{ss}\}$, which is fully determined by the pair of controllable variables ω_T, the transmon's frequency, and E_d, amplitude field of the pumping applied to the cavity. In order to implement the cycle, the stationary state is slowly varied (quasi-static) (The timescale for which the changes made can be considered slow is such that the conditions imposed to the system state are satisfied, namely the state is a product state and the cavity steady state is a coherent state with Equation (4)) by changing the "knobs" (ω_T, E_d). It is composed of four strokes where we keep one of the two controllable variables constant and vary the other one, for example, at the first stroke we keep $E_d = E_0$ and vary ω_T from ω_0 to ω_1. The complete cycle is sketched in Figure 2.

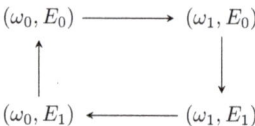

Figure 2. Sketch of the thermodynamic cycle obtained by varying the tunable parameters ω_T and E_d. Each one of the strokes are obtained by keeping one of the variables constant while quasi-statically varying the other one.

Since we are interested in analyzing the engine as a function of its parameters of operation, we simulated several cycles with boundary values (ω_1, E_1), which will range from the minimum value (ω_0, E_0) to the maximum one $(\omega_{1,max}, E_{1,max})$. The corresponding cycles lie on the von Neumann entropy surface depicted in Figure 3. In Appendix A plots of the stationary state's population and quantum coherence ρ_T^{ee} and $|\rho_T^{eg}|$ as a function of (ω_T, E_d) are shown. There we can observed clearly that the WS exhibits quantum coherence and population changes during its operation.

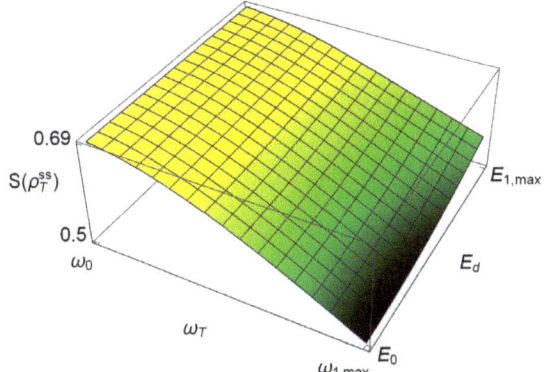

Figure 3. Stationary state's von Neumann entropy in the regime of operation of the thermal engine. Any thermodynamic cycle must be contained on this surface.

Finally, it is worth emphasizing that in the present analysis all parameters were chosen from an *experimentally accessible* regime [28,30–32], under the validity of the approximation of weak-coupling interaction between transmon and cavity. The parameters are collected in Table 1.

Table 1. Engine parameters used in the present analysis.

Parameter	Value
$\omega_{CPW}/2\pi$	4.94 GHz
$\omega/2\pi$	4.94 GHz
$g/2\pi\hbar$	120 MHz
T	30 mK
$\Gamma/2\pi$	2 MHz
$\kappa_{CPW}/2\pi$	1 MHz
$\omega_0/2\pi$	100 MHz
$\omega_{1,\max}/2\pi$	1000 MHz
$E_0/2\pi\hbar$	0.2 MHz
$E_{1,\max}/2\pi\hbar$	2 MHz

4. Work, Heat and Efficiency

The first law of thermodynamics, $\Delta E(t) = W(t) + Q(t)$, states that a variation of the internal energy along a thermodynamic process can be divided into two different parts, work $W(t)$ and heat $Q(t)$, where for Lindblad dynamics we have [4,37],

$$W(t) = \int_0^t \mathrm{tr}\left\{\rho(t')\dot{H}(t')dt'\right\},$$
$$Q(t) = \int_0^t \mathrm{tr}\left\{\dot{\rho}(t')H(t')dt'\right\}. \tag{8}$$

Typically, work is understood as a controllable energy exchange, which can be used for something useful, while heat cannot be controlled, emerging from the unavoidable interaction of the engine with its environment. As stated before, there are certain situations in which it can be shown that part of $Q(t)$ does not cause any entropic variation [24]. This has led to proposals for the differentiation of two distinct forms of energy contributions to Q: the *passive energy* $\mathcal{Q}(t)$, which is responsible for the variation in entropy, and the variation in *ergotropy* $\Delta \mathcal{W}(t)$ which is a "work-like energy" that can be extracted by means of a unitary transformation and consequently would not cause any entropic change. Both terms are defined as,

$$\mathcal{Q}(t) = \int_0^t \text{tr}\left\{\dot{\pi}(t')H(t')dt'\right\},$$
$$\Delta\mathcal{W}(t) = \int_0^t \text{tr}\left\{[\dot{\rho}(t') - \dot{\pi}(t')]H(t')dt'\right\}, \tag{9}$$

with $\pi(t)$ being the passive state [25] associated with the state $\rho(t)$ at time t. To calculate the upper bound on the efficiency for systems that exhibit these different "flavors" of energy one should replace Q by $\mathcal{Q}$ in statements of the second law, since the *ergotropy* is essentially a mechanical type of energy, and consequently not limited by the second law, resulting in a different upper bound, see also Ref. [24].

Distinguishing these types of energy exchanged with the environment is crucial when one is interested in determining the fundamental upper bounds on the efficiency. However, in the present context, we are more interested in experimentally relevant statements, i.e., computing the efficiency in terms of what can be measured directly. Thus, we consider the ratio of the extracted work to the total energy acquired from the bath, independent of its type [24].

The cycle designed here is such that in each stroke one of the knobs (ω_T, E_d) is kept fixed, while the other one is changed. Recall that the cavity is assumed to be a subpart of the bath seen by the WS, and that its state is modified by E_d. Since the WS is always in contact with the environment, one has that heat and work are exchanged in each stroke. Here, such a calculation is done by using Equation (8), considering the stationary state Equation (7) and the effective WS Hamiltonian Equation (6). Then, for the ith stroke, the corresponding W_i and Q_i integrals, representing the work and heat delivered (extracted) to (from) the WS, can be parametrized in terms of the respective knob variation as we can see in Appendix B. These quantities are obtained using the WS effective Hamiltonian $\hat{H}_{T,RWA}$, which already takes into account the interaction with the external bath and pumped cavity.

Once these quantities are determined, we can calculate the efficiency η of this engine, defined by

$$\eta = -\frac{\sum_{i=1}^4 W_i}{Q_+}, \tag{10}$$

with the delivered heat to the WS in a complete cycle being given by $Q_+ = \sum_{i=1}^4 Q_+^i$, with Q_+^i the given heat (only positive contributions inside the stroke) during the i_{th} stroke (see Appendix B). Therefore, this efficiency represents the amount of work extracted from the engine through the use of the delivered heat to the WS.

Figure 4 shows the engine efficiency η attained in the execution of the strokes as a function of the boundary values (ω_1, E_1), as depicted in Figure 2. Please note that (ω_1, E_1) sweeps the entire spectrum of the tunable parameters (ω_T, E_d), going from (ω_0, E_0) to $(\omega_{1,\max}, E_{1,\max})$ where we find the maximal efficiency. It is worth mentioning here that the highest value of the efficiency is dependent on the chosen regime of parameters, which in our case is based on experimentally attainable values [28,30–32]. As usual, in order to extract the predicted work, one has to couple our engine to another system. We envision using the experimental setup of Ref. [28], where a mechanical nanoresonator is present and weakly driven by the transmon. Thus, under such a configuration, by following the nanoresonator's state (recall that we have assumed infinite inertia, i.e., the transmon is not capable of changing the cavity's state. In situations where such an assumption does not hold, one has to take into account the possibility of having the transmon doing work on the cavity), one can determine the amount of energy transferred in the form of work. In addition, by observing the transmon's state, one can obtain the amount of heat given by the non-standard bath, providing a full characterization of our engine.

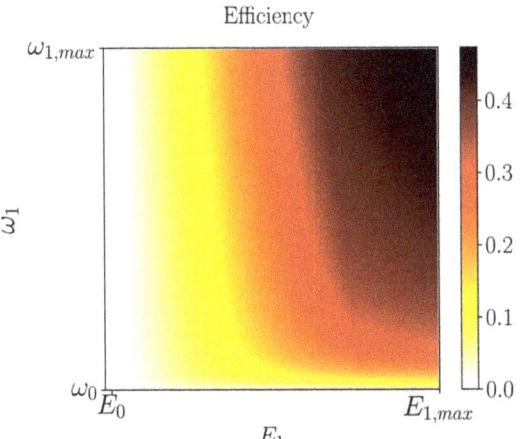

Figure 4. Efficiency η as a function of the upper values (ω_1, E_1) for the cycle depicted in Figure 2. The observed highest efficiency of about 47% was attained when $(\omega_1, E_1) = (\omega_{1,max}, E_{1,max})$, with $\omega_{1,max}/2\pi = 1000$ MHz and $E_{1,max}/2\pi\hbar = 2$ MHz.

5. Conclusions and Final Remarks

Theoretical research of small heat engines in the quantum domain is common place in quantum thermodynamics [37–46]. In the present work, we have devised a transmon-based heat engine using an experimentally realistic regime of parameters reaching a maximal efficiency of 47%, which turns out to be a reasonable value when compared with the state of the art in quantum heat engines. One of the most recent experiments in quantum heat engine was implemented by Peterson et al. [47] using a spin $-1/2$ system and nuclear resonance techniques, performing an Otto cycle with efficiency in excess of 42% at maximum power. It is important to stress that implementing small heat engines constitutes a hard task, even when dealing with classical systems. Indeed, a representative example is the single ion confined in a linear Paul trap with a tapered geometry, which was used to implement a Stirling engine [48] with efficiency of only 0.28%. Additional research is being carried out concerning the behavior of this engine influenced by the presence of coherence and the dimension of the WS. By devising this theoretical protocol for the implementation of a quantum engine, we hope to help the community, and in particular experimentalists, in the formidable task to design and implement quantum thermodynamic systems and to consolidate the concepts of this new exiting field of research.

Author Contributions: C.C., F.B. and S.D. equally contributed to conceptualization, investigation and writing of the paper.

Funding: C.C. and F.B. acknowledge financial support in part by the Coordenação de Aperfeiçoamento de Pessoal de Nível Superior—Brasil (CAPES)—Finance Code 001. During his stay at UMBC, C.C. was supported by the CAPES scholarship PDSE/process No. 88881.132982/2016-01. F.B. is also supported by the Brazilian National Institute for Science and Technology of Quantum Information (INCT-IQ) under Grant No. 465469/2014-0/CNPq. S.D. acknowledges support from the U.S. National Science Foundation under Grant No. CHE-1648973.

Acknowledgments: We thank F. Rouxinol and V. F. Teizen for valuable discussions. C.C. would like to thank the hospitality of UMBC, where most of this research was conducted.

Conflicts of Interest: The authors declare no conflict of interest.

Appendix A. Non-Thermal Equilibrium States

Here, we summarize the explicit expressions of the density matrix elements of ρ_T^{ss} (7), which are plotted as a function of (ω_T, E_d) in Figure A1.

$$\rho_T^{ee} = \frac{\frac{g^2 E_d^2}{\hbar^4\left[\frac{1}{4}\kappa_{CPW}^2 + (\omega_{CPW} - \omega)^2\right]} + \frac{1}{1+e^{\beta\hbar\omega_T}}\left[\frac{1}{4}\frac{\Gamma^2}{\tanh^2(\beta\hbar\omega_T/2)} + (\omega_T - \omega)^2\right]}{\frac{2g^2 E_d^2}{\hbar^4\left[\frac{1}{4}\kappa_{CPW}^2 + (\omega_{CPW} - \omega)^2\right]} + \left[\frac{1}{4}\frac{\Gamma^2}{\tanh^2(\beta\hbar\omega_T/2)} + (\omega_T - \omega)^2\right]}, \tag{A1}$$

$$\rho_T^{gg} = \frac{\frac{g^2 E_d^2}{\hbar^4\left[\frac{1}{4}\kappa_{CPW}^2 + (\omega_{CPW} - \omega)^2\right]} + \frac{1}{1+e^{-\beta\hbar\omega_T}}\left[\frac{1}{4}\frac{\Gamma^2}{\tanh^2(\beta\hbar\omega_T/2)} + (\omega_T - \omega)^2\right]}{\frac{2g^2 E_d^2}{\hbar^4\left[\frac{1}{4}\kappa_{CPW}^2 + (\omega_{CPW} - \omega)^2\right]} + \left[\frac{1}{4}\frac{\Gamma^2}{\tanh^2(\beta\hbar\omega_T/2)} + (\omega_T - \omega)^2\right]}, \tag{A2}$$

$$\rho_T^{eg} = \frac{\frac{1}{2\hbar}\left[\frac{\Gamma}{\tanh(\beta\hbar\omega_T/2)}i + 2(\omega_T - \omega)\right]}{\frac{2g^2 E_d^2}{\hbar^4\left[\frac{1}{4}\kappa_{CPW}^2 + (\omega_{CPW} - \omega)^2\right]} + \left[\frac{1}{4}\frac{\Gamma^2}{\tanh^2(\beta\hbar\omega_T/2)} + (\omega_T - \omega)^2\right]} \frac{gE_d}{\hbar\left[i\frac{\kappa_{CPW}}{2} - (\omega_{CPW} - \omega)\right]} \tanh(\beta\hbar\omega_T/2). \tag{A3}$$

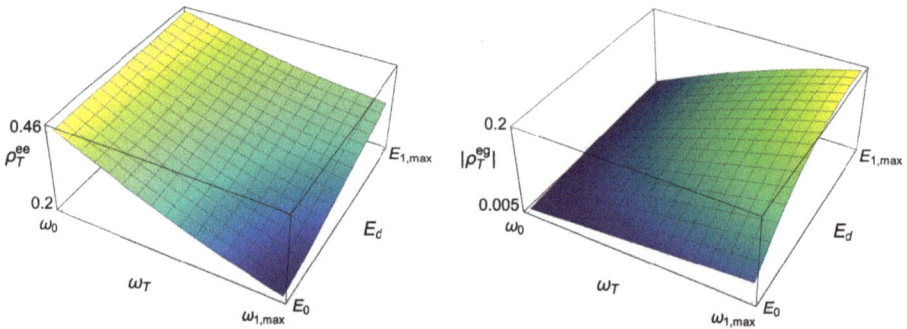

Figure A1. Stationary state's elements ρ_T^{ee} and $|\rho_T^{eg}|$ for different values of (ω_T, E_d). Important amounts of population and quantum coherence changes can be reached during the engine operation.

Appendix B. Thermodynamic Quantities along Each Stroke

In this appendix we summarize the explicit expressions of the thermodynamic quantities W_i and Q_i for $i = 1, 2, 3, 4$ and the heat Q_+ given to the WS. These quantities are obtained by changing quasi-statically the parameters ω_T and E_d producing a succession of steady states $\hat{\rho}_T^{ss}(\omega_T, E_d)$:

$$\begin{aligned}
W_1 &= \int_{\omega_0}^{\omega_1} \operatorname{tr}\left\{\hat{\rho}_T^{ss}(\omega_T, E_0)\left(\frac{\partial \hat{H}_{T,RWA}}{\partial \omega_T}\right)_{E_0}\right\} d\omega_T, \\
W_2 &= \int_{E_0}^{E_1} \operatorname{tr}\left\{\hat{\rho}_T^{ss}(\omega_1, E_d)\left(\frac{\partial \hat{H}_{T,RWA}}{\partial E_d}\right)_{\omega_1}\right\} dE_d, \\
W_3 &= \int_{\omega_1}^{\omega_0} \operatorname{tr}\left\{\hat{\rho}_T^{ss}(\omega_T, E_1)\left(\frac{\partial \hat{H}_{T,RWA}}{\partial \omega_T}\right)_{E_1}\right\} d\omega_T, \\
W_4 &= \int_{E_1}^{E_0} \operatorname{tr}\left\{\hat{\rho}_T^{ss}(\omega_0, E_d)\left(\frac{\partial \hat{H}_{T,RWA}}{\partial E_d}\right)_{\omega_0}\right\} dE_d.
\end{aligned} \tag{A4}$$

$$Q_1 = \int_{\omega_0}^{\omega_1} \text{tr}\left\{\left(\frac{\partial \rho_T^{ss}}{\partial \omega_T}\right)_{E_0} \tilde{H}_{T,\text{RWA}}(\omega_T, E_0)\right\} d\omega_T,$$

$$Q_2 = \int_{E_0}^{E_1} \text{tr}\left\{\left(\frac{\partial \rho_T^{ss}}{\partial E_d}\right)_{\omega_1} \tilde{H}_{T,\text{RWA}}(\omega_1, E_d)\right\} dE_d,$$

$$Q_3 = \int_{\omega_1}^{\omega_0} \text{tr}\left\{\left(\frac{\partial \rho_T^{ss}}{\partial \omega_T}\right)_{E_1} \tilde{H}_{T,\text{RWA}}(\omega_T, E_1)\right\} d\omega_T,$$

$$Q_4 = \int_{E_1}^{E_0} \text{tr}\left\{\left(\frac{\partial \rho_T^{ss}}{\partial E_d}\right)_{\omega_0} \tilde{H}_{T,\text{RWA}}(\omega_0, E_d)\right\} dE_d. \quad \text{(A5)}$$

$$Q_+ = \sum_{i=1}^{4} Q_+^i \quad \text{(A6)}$$

with Q_+^i for $i = 1, 2, 3, 4$ given by

$$Q_+^1 = \int_{\omega_0}^{\omega_1} \text{tr}\left\{\left(\frac{\partial \rho_T^{ss}}{\partial \omega_T}\right)_{E_0} \tilde{H}_{T,\text{RWA}}(\omega_T, E_0)\right\} \Theta\left[\text{tr}\left\{\left(\frac{\partial \rho_T^{ss}}{\partial \omega_T}\right)_{E_0} \tilde{H}_{T,\text{RWA}}(\omega_T, E_0)\right\}\right] d\omega_T,$$

$$Q_+^2 = \int_{E_0}^{E_1} \text{tr}\left\{\left(\frac{\partial \rho_T^{ss}}{\partial E_d}\right)_{\omega_1} \tilde{H}_{T,\text{RWA}}(\omega_1, E_d)\right\} \Theta\left[\text{tr}\left\{\left(\frac{\partial \rho_T^{ss}}{\partial E_d}\right)_{\omega_1} \tilde{H}_{T,\text{RWA}}(\omega_1, E_d)\right\}\right] dE_d,$$

$$Q_+^3 = \int_{\omega_1}^{\omega_0} \text{tr}\left\{\left(\frac{\partial \rho_T^{ss}}{\partial \omega_T}\right)_{E_1} \tilde{H}_{T,\text{RWA}}(\omega_T, E_1)\right\} \Theta\left[\text{tr}\left\{\left(\frac{\partial \rho_T^{ss}}{\partial \omega_T}\right)_{E_1} \tilde{H}_{T,\text{RWA}}(\omega_T, E_1)\right\}\right] d\omega_T,$$

$$Q_+^4 = \int_{E_1}^{E_0} \text{tr}\left\{\left(\frac{\partial \rho_T^{ss}}{\partial E_d}\right)_{\omega_0} \tilde{H}_{T,\text{RWA}}(\omega_0, E_d)\right\} \Theta\left[\text{tr}\left\{\left(\frac{\partial \rho_T^{ss}}{\partial E_d}\right)_{\omega_0} \tilde{H}_{T,\text{RWA}}(\omega_0, E_d)\right\}\right] dE_d. \quad \text{(A7)}$$

where the Heaviside function $\Theta[\cdot]$ is inside the integral, selecting only the positive contributions (heat given to the WS) along the stroke.

References

1. Gemmer, J.; Michel, M.; Mahler, G. *Quantum Thermodynamics*; Springer: Berlin/Heidelberg, Germany, 2004.
2. Gelin, M.F.; Thoss, M. Thermodynamics of a subensemble of a canonical ensemble. *Phys. Rev. E* **2009**, *79*, 051121. [CrossRef]
3. Scully, M.O.; Zubairy, M.S.; Agarwal, G.S.; Walther, H. Extracting Work from a Single Heat Bath via Vanishing Quantum Coherence. *Science* **2003**, *299*, 862. [CrossRef] [PubMed]
4. Gardas, B.; Deffner, S. Thermodynamic universality of quantum Carnot engines. *Phys. Rev. E* **2015**, *92*, 042126. [CrossRef] [PubMed]
5. Deffner, S. Efficiency of Harmonic Quantum Otto Engines at Maximal Power. *Entropy* **2018**, *20*, 875. [CrossRef]
6. Çakmak, B.; Müstecaplıoğlu, O.E. Spin quantum heat engines with shortcuts to adiabaticity. *Phys. Rev. E* **2019**, *99*, 032108. [CrossRef]
7. Klaers, J.; Faelt, S.; Imamoglu, A.; Togan, E. Squeezed Thermal Reservoirs as a Resource for a Nanomechanical Engine beyond the Carnot Limit. *Phys. Rev. X* **2017**, *7*, 031044. [CrossRef]
8. Dillenschneider, R.; Lutz, E. Energetics of quantum correlations. *EPL (Europhys. Lett.)* **2009**, *88*, 50003. [CrossRef]
9. Huang, X.L.; Wang, T.; Yi, X.X. Effects of reservoir squeezing on quantum systems and work extraction. *Phys. Rev. E* **2012**, *86*, 051105. [CrossRef] [PubMed]
10. Abah, O.; Lutz, E. Efficiency of heat engines coupled to nonequilibrium reservoirs. *EPL (Europhys. Lett.)* **2014**, *106*, 20001. [CrossRef]
11. Roßnagel, J.; Abah, O.; Schmidt-Kaler, F.; Singer, K.; Lutz, E. Nanoscale Heat Engine Beyond the Carnot Limit. *Phys. Rev. Lett.* **2014**, *112*, 030602. [CrossRef]
12. Hardal, A.Ü.C.; Müstecaplıoğlu, Ö.E. Superradiant Quantum Heat Engine. *Sci. Rep.* **2015**, *5*, 12953. [CrossRef]
13. Niedenzu, W.; Gelbwaser-Klimovsky, D.; Kofman, A.G.; Kurizki, G. On the operation of machines powered by quantum non-thermal baths. *New J. Phys.* **2016**, *18*, 083012. [CrossRef]

14. Manzano, G.; Galve, F.; Zambrini, R.; Parrondo, J.M.R. Entropy production and thermodynamic power of the squeezed thermal reservoir. *Phys. Rev. E* **2016**, *93*, 052120. [CrossRef]
15. Agarwalla, B.K.; Jiang, J.H.; Segal, D. Quantum efficiency bound for continuous heat engines coupled to noncanonical reservoirs. *Phys. Rev. B* **2017**, *96*, 104304. [CrossRef]
16. Stefanatos, D. Optimal efficiency of a noisy quantum heat engine. *Phys. Rev. E* **2014**, *90*, 012119. [CrossRef]
17. Torrontegui, E.; Kosloff, R. Quest for absolute zero in the presence of external noise. *Phys. Rev. E* **2013**, *88*, 032103. [CrossRef]
18. Gardas, B.; Deffner, S.; Saxena, A. Non-hermitian quantum thermodynamics. *Sci. Rep.* **2016**, *6*, 23408. [CrossRef]
19. Hatano, T.; Sasa, S.I. Steady-State Thermodynamics of Langevin Systems. *Phys. Rev. Lett.* **2001**, *86*, 3463. [CrossRef]
20. Oono, Y.; Paniconi, M. Steady State Thermodynamics. *Prog. Theor. Phys. Suppl.* **1998**, *130*, 29. [CrossRef]
21. Horowitz, J.M.; Sagawa, T. Equivalent Definitions of the Quantum Nonadiabatic Entropy Production. *J. Stat. Phys.* **2014**, *156*, 55. [CrossRef]
22. Yuge, T.; Sagawa, T.; Sugita, A.; Hayakawa, H. Geometrical Excess Entropy Production in Nonequilibrium Quantum Systems. *J. Stat. Phys.* **2013**, *153*, 412. [CrossRef]
23. Binder, F.; Vinjanampathy, S.; Modi, K.; Goold, J. Quantum thermodynamics of general quantum processes. *Phys. Rev. E* **2015**, *91*, 032119. [CrossRef]
24. Niedenzu, W.; Mukherjee, V.; Ghosh, A.; Kofman, A.G.; Kurizki, G. Quantum engine efficiency bound beyond the second law of thermodynamics. *Nat. Commun.* **2018**, *9*, 165. [CrossRef]
25. Pusz, W.; Woronowicz, S.L. Passive states and KMS states for general quantum systems. *Commun. Math. Phys.* **1978**, *58*, 273. [CrossRef]
26. Allahverdyan, A.E.; Balian, R.; Nieuwenhuizen, T.M. Maximal work extraction from finite quantum systems. *EPL (Europhys. Lett.)* **2004**, *67*, 565. [CrossRef]
27. Manzano, G. Squeezed thermal reservoir as a generalized equilibrium reservoir. *Phys. Rev. E* **2018**, *98*, 042123. [CrossRef]
28. Rouxinol, F.; Hao, Y.; Brito, F.; Caldeira, A.O.; Irish, E.K.; LaHaye, M.D. Measurements of nanoresonator-qubit interactions in a hybrid quantum electromechanical system. *Nanotechnology* **2016**, *27*, 364003. [CrossRef]
29. Callen, H.B. *Thermodynamics and an Introduction to Thermostatistics*; Wiley: Hoboken, NJ, USA, 1985.
30. Kok, P.; Munro, W.J.; Nemoto, K.; Ralph, T.C.; Dowling, J.P.; Milburn, G.J. Linear optical quantum computing with photonic qubits. *Rev. Mod. Phys.* **2007**, *79*, 135. [CrossRef]
31. Hofheinz, M.; Weig, E.M.; Ansmann, M.; Bialczak, R.C.; Lucero, E.; Neeley, M.; O'Connell, A.D.; Wang, H.; Martinis, J.M.; Cleland, A.N. Generation of Fock states in a superconducting quantum circuit. *Nature* **2008**, *454*, 310. [CrossRef]
32. Mallet, F.; Ong, F.R.; Palacios-Laloy, A.; Nguyen, F.; Bertet, P.; Vion, D.; Esteve, D. Single-shot qubit readout in circuit Quantum Electrodynamics. *Nat. Phys.* **2009**, *5*, 791. [CrossRef]
33. Majer, J.; Chow, J.M.; Gambetta, J.M.; Koch, J.; Johnson, B.R.; Schreier, J.A.; Frunzio, L.; Schuster, D.I.; Houck, A.A.; Wallraff, A.; et al. Coupling Superconducting Qubits via a Cavity Bus. *Nature* **2007**, *449*, 443. [CrossRef]
34. Scully, M.O.; Zubairy, M.S. *Quantum Optics*; Cambridge University Press: Cambridge, UK, 1997. [CrossRef]
35. Deffner, S.; Jarzynski, C. Information Processing and the Second Law of Thermodynamics: An Inclusive, Hamiltonian Approach. *Phys. Rev. X* **2013**, *3*, 041003. [CrossRef]
36. Sasa, S.I.; Tasaki, H. Steady State Thermodynamics. *J. Stat. Phys.* **2006**, *125*, 125. [CrossRef]
37. Alicki, R. The quantum open system as a model of the heat engine. *J. Phys. A Math. Gen.* **1979**, *12*, L103. [CrossRef]
38. Geva, E.; Kosloff, R. A quantum-mechanical heat engine operating in finite time. A model consisting of spin-1/2 systems as the working fluid. *J. Chem. Phys* **1998**, *96*, 3054. [CrossRef]
39. Kieu, T.D. The Second Law, Maxwell's Demon, and Work Derivable from Quantum Heat Engines. *Phys. Rev. Lett.* **2004**, *93*, 140403. [CrossRef] [PubMed]
40. Quan, H.T.; Liu, Y.X.; Sun, C.P.; Nori, F. Quantum thermodynamic cycles and quantum heat engines. *Phys. Rev. E* **2007**, *76*, 031105. [CrossRef]
41. Linden, N.; Popescu, S.; Skrzypczyk, P. How Small Can Thermal Machines Be? The Smallest Possible Refrigerator. *Phys. Rev. Lett.* **2010**, *105*, 130401. [CrossRef]

42. Correa, L.A.; Palao, J.P.; Alonso, D.; Adesso, G. Quantum-enhanced absorption refrigerators. *Sci. Rep.* **2014**, *4*, 3949. [CrossRef]
43. Uzdin, R.; Levy, A.; Kosloff, R. Equivalence of Quantum Heat Machines, and Quantum-Thermodynamic Signatures. *Phys. Rev. X* **2015**, *5*, 031044. [CrossRef]
44. Abah, O.; Roßnagel, J.; Jacob, G.; Deffner, S.; Schmidt-Kaler, F.; Singer, K.; Lutz, E. Single-Ion Heat Engine at Maximum Power. *Phys. Rev. Lett.* **2012**, *109*, 203006. [CrossRef]
45. Zhang, K.; Bariani, F.; Meystre, P. Quantum Optomechanical Heat Engine. *Phys. Rev. Lett.* **2014**, *112*, 150602. [CrossRef]
46. Dawkins, S.T.; Abah, O.; Singer, K.; Deffner, S. Single Atom Heat Engine in a Tapered Ion Trap. In *Thermodynamics in the Quantum Regime: Fundamental Aspects and New Directions*; Binder, F., Correa, L.A., Gogolin, C., Anders, J., Adesso, G., Eds.; Springer International Publishing: Cham, Switzerland, 2018; pp. 887–896. [CrossRef]
47. Peterson, J.P.S.; Batalhão, T.B.; Herrera, M.; Souza, A.M.; Sarthour, R.S.; Oliveira, I.S.; Serra, R.M. Experimental characterization of a spin quantum heat engine. *arXiv* **2018**, arXiv:1803.06021.
48. Roßnagel, J.; Dawkins, S.T.; Tolazzi, K.N.; Abah, O.; Lutz, E.; Schmidt-Kaler, F.; Singer, K. A single-atom heat engine. *Science* **2016**, *352*, 325. [CrossRef]

© 2019 by the authors. Licensee MDPI, Basel, Switzerland. This article is an open access article distributed under the terms and conditions of the Creative Commons Attribution (CC BY) license (http://creativecommons.org/licenses/by/4.0/).

Article
A Quantum Heat Exchanger for Nanotechnology

Amjad Aljaloud [1,2,*], Sally A. Peyman [1,3] and Almut Beige [1]

1. The School of Physics and Astronomy, University of Leeds, Leeds LS2 9JT, UK;
   S.Peyman@leeds.ac.uk (S.A.P.); a.beige@leeds.ac.uk (A.B.)
2. Department of Physics, College of Sciences, University of Hail, Hail PO Box 2440, Saudi Arabia
3. Leeds Institute for Medical Research, School of Medicine, Wellcome Trust Brenner Building,
   St James' Teaching Hospital, University of Leeds, Leeds LS9 7TF, UK
* Correspondence: ml17asma@leeds.ac.uk

Received: 6 February 2020; Accepted: 21 March 2020; Published: 26 March 2020

Abstract: In this paper, we design a quantum heat exchanger which converts heat into light on relatively short quantum optical time scales. Our scheme takes advantage of heat transfer as well as collective cavity-mediated laser cooling of an atomic gas inside a cavitating bubble. Laser cooling routinely transfers individually trapped ions to nano-Kelvin temperatures for applications in quantum technology. The quantum heat exchanger which we propose here might be able to provide cooling rates of the order of Kelvin temperatures per millisecond and is expected to find applications in micro- and nanotechnology.

Keywords: quantum thermodynamics; laser cooling; cavitation; sonoluminescence

---

1. Introduction

Since its discovery in 1975 [1,2], laser cooling of individually trapped atomic particles has become a standard technique in quantum optics laboratories worldwide [3,4]. Rapidly oscillating electric fields can be used to strongly confine charged particles, such as single ions, for relatively large amounts of time [5]. Moreover, laser trapping provides unique means to control the dynamics of neutral particles, such as neutral atoms [6,7]. To cool single atomic particles, laser fields are applied which remove vibrational energy at high enough rates to transfer them down to near absolute-zero temperatures [5]. Nowadays, ion traps are used to perform a wide range of high-precision quantum optics experiments. For example, individually trapped ions are at the heart of devices with applications in quantum technology, such as atomic and optical clocks [8,9], quantum computers [10–13], quantum simulators [14,15] and electric and magnetic field sensors [16].

For laser cooling to be at its most efficient, the confinement of individually trapped particles should be so strong that the quantum characteristics of their motion is no longer negligible. This means that their vibrational energy is made up of energy quanta, which have been named phonons. When this applies, an externally applied laser field not only affects the electronic states of a trapped ion, but it also changes its vibrational state. Ideally, laser frequencies should be chosen such that the excitation of the ion should be most likely accompanied by the loss of a phonon. If the ion returns subsequently into its ground state via the spontaneous emission of a photon, its phonon state remains the same. Overall one phonon is permanently lost from the system which implies cooling. On average, every emitted photon lowers the vibrational energy of the trapped ion by the energy of one phonon. Eventually, the cooling process stops when the ion no longer possesses any vibrational energy.

Currently, there are many different ways of designing and fabricating ion traps [17,18]. However, the main requirements for the efficient conversion of vibrational energy into light on relatively short quantum optical time scales are always the same [19,20]:

(1) Individual atomic particles need to be so strongly confined that the quantum character of their motion has to be taken into account. In the following, ν denotes the phonon frequency and $\hbar\nu$ is the energy of a single phonon.
(2) A laser field with a frequency ω_L below the atomic transition frequency ω_0 needs to be applied. As long as the laser detuning $\Delta = \omega_0 - \omega_L$ and the phonon frequency ν are comparable in size,

$$\Delta \sim \nu, \tag{1}$$

the excitation of an ion is more likely accompanied by the annihilation of a phonon than by the creation of a phonon. Transitions which result in the simultaneous excitation of an ion and the creation of a phonon are possible but are less likely to occur as long as their detuning is larger.
(3) When excited, the confined atomic particle needs to be able to emit a photon. In the following, we denote its spontaneous decay rate by Γ. This rate should not be much larger than ν,

$$\nu \geq \Gamma, \tag{2}$$

so that the cooling laser couples efficiently to atomic transitions. At the same time, Γ should not be too small so that de-excitation of the excited atomic state happens often via the spontaneous emission of a photon.

Given these three conditions, the applied laser field results in the conversion of the vibrational energy of individually trapped ions into photons. As mentioned already above, laser cooling can prepare individually trapped atomic particles at low enough temperatures for applications in high-precision quantum optics experiments and in quantum technology.

In this paper, we ask the question whether laser cooling could also have applications in micro- and nanoscale physics experiments. For example, nanotechnology deals with objects which have dimensions between 1 and 1000 nm and is well known for its applications in information and communication technology, as well as sensing and imaging. Increasing the speed at which information can be processed and the sensitivity of sensors is usually achieved by reducing system dimensions. However, smaller devices are usually more prone to heating as thermal resistances increase [21]. Sometimes, large surface to volume ratios can help to off-set this problem. Another problem for nanoscale sensors is thermal noise. As sensors are reduced in size, their signal to noise ratio usually decreases and thus the thermal energy of the system can limit device sensitivity [22]. Therefore, thermal considerations have to be taken into account and large vacuums or compact heat exchangers have already become an integral part of nanotechnology devices.

Usually, heat exchangers in micro- and nanotechnology rely on fluid flow [23]. In this paper, we propose an alternative approach. More concretely, we propose to use heat transfer as well as a variation of laser cooling, namely cavity-mediated collective laser cooling [24–28], to lower the temperature of a small device. As illustrated in Figure 1, the proposed quantum heat exchanger mainly consist of a liquid which contains a large number of cavitating bubbles filled with noble gas atoms. Transducers constantly change the radius of these bubbles which should resemble optical cavities when they reach their minimum radius during bubble collapse phases. At this point, a continuously applied external laser field rapidly transfers vibrational energy of the atoms into light. If the surrounding liquid contains many cavitating bubbles, their surface area becomes relatively large and there can be a very efficient exchange of heat between the inside and the outside of cavitating bubbles. Any removal of thermal energy from the trapped atomic gas inside bubbles should eventually result in the cooling of the surrounding liquid and of the surface area of the device on which it is placed.

In this paper, we emphasize that cavitating bubbles can provide all of the above listed requirements for laser cooling, especially a very strong confinement of atomic particles, such as nitrogen [29,30]. For example, calculations based on a variation of the Rayleigh–Plesset equations show that the pressure at the location of a cavitating bubble can be significantly larger than the externally applied driving pressure [31]. However, the strongest indication for the presence of phonon modes with sufficiently large frequencies for laser cooling to work comes from the fact that sonoluminescence experiments are well-known for converting sound into relatively large amounts of thermal energy, while producing light in the optical regime [32–34]. During this process, the atomic gas inside a cavitating bubbles can reach very high temperatures [35,36], which hints at very strong couplings between electronic and vibrational degrees of freedom. In addition, the surfaces of cavitating bubbles can become opaque during the bubble collapse phase [37], thereby creating a spherical optical cavity [38,39] which is an essential requirement for cavity-mediated collective laser cooling.

To initiate the cooling process, an appropriately detuned laser field needs to be applied in addition to the transducers which confine the bubbles with sound waves. Although sonoluminescence has been studied in great detail and the idea of applying laser fields to cavitating bubbles is not new [40], not enough is known about the relevant quantum properties, such as phonon frequencies. Hence, we cannot predict realistic cooling rates for the experimental setup shown in Figure 1. A crude estimate which borrows data from different, already available experiments suggests that it might be possible to achieve cooling rates of the order of Kelvin temperatures per millisecond for volumes of liquid on a cubic micrometer scale. Cavitating bubbles already have applications in sonochemistry, where they are used to provide energy for chemical reactions [41]. Here, we propose to exploit the atom–phonon interactions in sonoluminescence experiments for laser cooling. In the presence of an appropriately detuned laser field, we expect other, highly-detuned heating processes to become secondary.

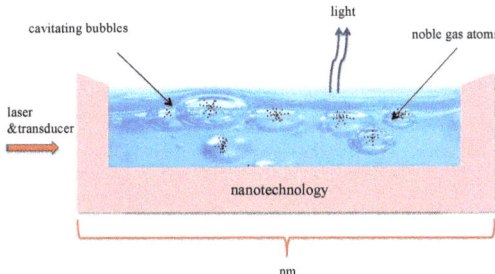

Figure 1. Schematic view of the proposed quantum heat exchanger. It consists of a liquid in close contact with the area which we want to cool. The liquid should contain cavitating bubbles which are filled with atomic particles, such as nitrogen, and should be driven by sounds waves and laser light. The purpose of the sound waves is to constantly change bubble sizes. The purpose of the laser is to convert thermal energy during bubble collapse phases into light.

There are five sections in this paper. The purpose of Section 2 is to provide an introduction to cavity-mediated collective laser cooling of an atomic gas. As we show below, this technique is a variation of standard laser cooling techniques for individually trapped atomic particles. We provide an overview of the experimental requirements and estimate achievable cooling rates. Section 3 studies the effect of thermalization for a large collection of atoms with elastic collisions. Section 4 reviews the main design principles of a quantum heat exchanger for nanotechnology. Finally, we summarize our findings in Section 5.

2. Cavity-Mediated Collective Laser Cooling

In this section, we first have a closer look at a standard laser cooling technique for an individually trapped atomic particle [19,20]. Afterwards, we review cavity-mediated laser cooling of a single atom [42–46] and of an atomic gas [26–28].

2.1. Laser Cooling of Individually Trapped Particles

Figure 2a shows a single two-level atom (or ion) with external laser driving inside an approximately harmonic trapping potential. Most importantly, the atom should be so strongly confined that its phonon states are no longer negligible. In the following, ν denotes the frequency of the energy quanta in the vibrational energy of the atomic particle and $|m\rangle$ is a vibrational state with exactly m phonons. Moreover, $|g\rangle$ and $|e\rangle$ denote the ground and the excited electronic state of the trapped particle with energy separation $\hbar\omega_0$. Figure 2b shows the energy level of the combined atom–phonon system with the energy eigenstates $|x, m\rangle$.

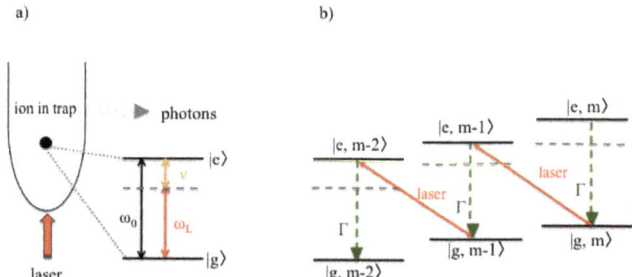

Figure 2. (a) Schematic view of the experimental setup for laser cooling of a single trapped ion. Here, $|g\rangle$ and $|e\rangle$ denote the ground and the excited state of the ion, respectively, with transition frequency ω_0 and spontaneous decay rate Γ. The motion of the particle is strongly confined by an external harmonic trapping potential such that it quantum nature can no longer be neglected. Here, ν denotes the frequency of the corresponding phonon mode and ω_L is the frequency of the applied cooling laser. (b) The purpose of the laser is to excite the ion, while annihilating a phonon, thereby causing transitions between the basis states $|x, m\rangle$ with x = g, e and m = 0, 1, ... of the atom–phonon system. If the excitation of the ion is followed by the spontaneous emission of a photon, a phonon is permanently lost, which implies cooling.

To lower the temperature of the atom, the frequency ω_L of the cooling laser needs to be below its transition frequency ω_0. Ideally, the laser detuning $\Delta = \omega_0 - \omega_L$ equals the phonon frequency ν (cf. Equation (1)). In addition, the spontaneous decay rate Γ of the excited atomic state should not exceed ν (cf. Equation (2)). When both conditions apply, the cooling laser couples most strongly, i.e., resonantly and efficiently, to transitions for which the excitation of the atom is accompanied by the simultaneously annihilation of a phonon. All other transitions are strongly detuned. Moreover, the spontaneous emission of a photon only affects the electronic but not the vibrational states of the atom. Hence, the spontaneous emission of a photon usually indicates the loss of one phonon. Suppose the atom was initially prepared in a state $|g, m\rangle$. Then, its final state equals $|g, m-1\rangle$. One phonon has been permanently removed from the system which implies cooling. As illustrated in Figure 2b, the trapped particle eventually reaches its ground state $|g, 0\rangle$ where it no longer experiences the cooling due to off-resonant driving [19,20].

To a very good approximation, the Hamiltonian of the atom–phonon system equals [20]

$$H_I = \hbar g \left(\sigma^- b^\dagger + \sigma^+ b \right) \quad (3)$$

in the interaction picture with respect to its free energy. Here, g denotes the (real) atom–phonon coupling constant, while $\sigma^+ = |e\rangle\langle g|$ and $\sigma^- = |g\rangle\langle e|$ are atomic rising and lowering operators. Moreover, b and $b^\dagger$ are phonon annihilation and creation operators with $[b, b^\dagger] = 1$. To take into account the spontaneous emission of photons from the excited state of the atom with decay rate Γ, we describe the atom–phonon system in the following by its density matrix $\rho_I(t)$ with

$$\dot{\rho}_I = -\frac{i}{\hbar}[H_I, \rho_I] + \Gamma\left(\sigma^- \rho_I \sigma^+ - \frac{1}{2}\sigma^+\sigma^- \rho_I - \frac{1}{2}\rho_I \sigma^+\sigma^-\right). \tag{4}$$

This equation can be used to analyze the dynamics of the expectation value $\langle A_I \rangle = \text{Tr}(A_I \rho_I)$ of observables A_I, since it implies

$$\langle \dot{A}_I \rangle = -\frac{i}{\hbar}\langle[A_I, H_I]\rangle + \Gamma\left\langle \sigma^+ A_I \sigma^- - \frac{1}{2} A_I \sigma^+\sigma^- - \frac{1}{2}\sigma^+\sigma^- A_I \right\rangle. \tag{5}$$

Here, we are especially interested in the dynamics of the mean phonon number $m = \langle b^\dagger b \rangle$. To obtain a closed set of rate equations, we also need to study the dynamics of the population of the excited atomic state $s = \langle \sigma^+ \sigma^- \rangle$ and the dynamics of the atom–phonon coherence $k_1 = i\langle \sigma^- b^\dagger - \sigma^+ b \rangle$. Using Equation (5), one can show that

$$\begin{aligned}
\dot{m} &= -g k_1, \\
\dot{s} &= g k_1 - \Gamma s, \\
\dot{k}_1 &= 2g(m-s) - 4g\, ms - \frac{1}{2}\Gamma k_1
\end{aligned} \tag{6}$$

when assuming that $\langle \sigma^+ \sigma^- b^\dagger b \rangle = \langle \sigma^+ \sigma^- \rangle \langle b^\dagger b \rangle = ms$ to a very good approximation. Having a closer look at the above equations, we see that the system rapidly reaches its stationary state with $m = s = k_1 = 0$. Eventually, the atom reaches a very low temperature. More detailed calculations reveal that the final phonon m of the trapped atom depends on its system parameters but remains small as long as the ratio Γ/ν is sufficiently small [20]. The above cooling equations (Equation (6)) also show that the corresponding cooling rate equals

$$\gamma^{\text{standard}}_{1\,\text{atom}} = g^2/\Gamma \tag{7}$$

to a very good approximation and that the cooling process takes place not on mechanical but on relatively short quantum optical time scales.

2.2. Cavity-Mediated Laser Cooling of a Single Atom

Suppose we want to cool a single atom whose transition frequency ω_0 is well above the optical regime, i.e., much larger than typical laser frequencies ω_L. In this case, it is impossible to realize the condition $\Delta \sim \nu$ in Equation (1). Hence, it might seem impossible to lower the temperature of the atom via laser cooling. To overcome this problem, we confine the particle in the following inside an optical resonator (cf. Figure 3) and denote the cavity state with exactly n photons by $|n\rangle$. Using this notation, the energy eigenstates of the atom–phonon–photon systems can be written as $|x, m, n\rangle$. Moreover, ν is again the phonon frequency, κ denotes the spontaneous cavity decay rate and ω_L and ω_{cav} denote the laser and the cavity frequency, respectively.

In the experimental setup in Figure 3, all transitions which result in the excitation of the atom are naturally strongly detuned and can be neglected. However, the same does not have to apply to indirect couplings which result in the direct conversion of phonons into cavity photons [27,44]. Suppose the cavity detuning $\Delta_{\text{cav}} = \omega_{\text{cav}} - \omega_L$ and the phonon frequency ν are approximately the same and the cavity decay rate κ does not exceed ν,

$$\Delta_{\text{cav}} \sim \nu \quad \text{and} \quad \nu \geq \kappa, \tag{8}$$

in analogy to Equations (1) and (2). Then, two-step transitions which excite the atom while annihilating a phonon immediately followed by the de-excitation of the atom while creating a cavity photon become resonant and dominate the dynamics of the atom–phonon–photon system. The overall effect of these two-step transitions is the direct conversion of a phonon into a cavity photon, while the atom remains essentially in its ground state (cf. Figure 3b). When a cavity photon subsequently leaks into the environment, the phonon is permanently lost.

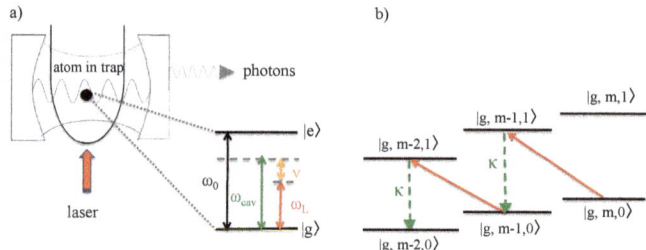

Figure 3. (a) Schematic view of the experimental setup for cavity-mediated laser cooling of a single atom. The main difference between this setup and the setup shown in Figure 2 is that the atom now couples in addition to an optical cavity with frequency ω_{cav} and the spontaneous decay rate κ. Here, both the cavity field and the laser are highly detuned from the atomic transition and the direct excitation of the atom remains negligible. However, the cavity detuning $\Delta_{\text{cav}} = \omega_{\text{cav}} - \omega_{\text{L}}$ should equal the phonon frequency of the trapped particle. (b) As a result, only the annihilation of a phonon accompanied by the simultaneous creation of a cavity photon are in resonance. In cavity-mediated laser cooling, the purpose of the laser is to convert phonons into cavity photons. The subsequent loss of this photon via spontaneous emission results in the permanent loss of a phonon and therefore in the cooling of the trapped particle.

To model the above described dynamics, we describe the experimental setup in Figure 3 in the following by the interaction Hamiltonian [44,45]

$$H_{\text{I}} = \hbar g_{\text{eff}} \left(bc^\dagger + b^\dagger c \right), \tag{9}$$

where g_{eff} denotes the effective atom–cavity coupling constant and where c with $[c, c^\dagger] = 1$ is the cavity photon annihilation operator. Since the atom remains essentially in its ground state, its spontaneous photon emission remains negligible. To model the possible leakage photons through the cavity mirrors, we employ again a master equation. Doing so, the time derivative of the density matrix $\rho_{\text{I}}(t)$ of the phonon–photon system equals

$$\dot{\rho}_{\text{I}} = -\frac{i}{\hbar}[H_{\text{I}}, \rho_{\text{I}}] + \kappa \left(c\rho_{\text{I}} c^\dagger - \frac{1}{2} c^\dagger c \rho_{\text{I}} - \frac{1}{2} \rho_{\text{I}} c^\dagger c \right) \tag{10}$$

in the interaction picture. Hence, expectation values $\langle A_{\text{I}} \rangle = \text{Tr}(A_{\text{I}} \rho_{\text{I}})$ of phonon–photon observables A_{I} evolve such that

$$\langle \dot{A}_{\text{I}} \rangle = -\frac{i}{\hbar}[A_{\text{I}}, H_{\text{I}}] + \kappa \left\langle c^\dagger A_{\text{I}} c - \frac{1}{2} A_{\text{I}} c^\dagger c - \frac{1}{2} c^\dagger c A_{\text{I}} \right\rangle, \tag{11}$$

in analogy to Equation (5). In the following, we use this equation to study the dynamics of the phonon number $m = \langle b^\dagger b \rangle$, the photon number $n = \langle c^\dagger c \rangle$, and the phonon–photon coherence $k_1 = i\langle bc^\dagger - b^\dagger c \rangle$. Proceeding as described in the previous subsection, we now obtain the rate equations

$$\begin{aligned} \dot{m} &= g_{\text{eff}} k_1, \\ \dot{n} &= -g_{\text{eff}} k_1 - \kappa n, \\ \dot{k}_1 &= 2g_{\text{eff}}(n - m) - \frac{1}{2}\kappa k_1. \end{aligned} \quad (12)$$

These describe the continuous conversion of phonons into cavity photons which subsequently escape the system. Hence, it is not surprising to find that the stationary state of the atom–phonon–photon system corresponds to $m = n = k_1 = 0$. Independent of its initial state, the atom again reaches a very low temperature. In analogy to Equation (7), the effective cooling rate for cavity-mediated laser cooling is now given by [44,45]

$$\gamma_{1\,\text{atom}} = g_{\text{eff}}^2/\kappa. \quad (13)$$

Due to the resonant coupling being indirect, g_{eff} is in general a few orders of magnitude smaller than g in Equation (7), if the spontaneous decay rates κ and Γ are of similar size. Cooling a single atom inside an optical resonator might therefore take significantly longer. However, as we show below, this reduction in cooling rate can be compensated for by the collective enhancement of the atom–cavity interaction constant g_{eff} [26].

2.3. Cavity-Mediated Collective Laser Cooling of an Atomic Gas

Finally, we have a closer look at cavity-mediated collective laser cooling of an atomic gas inside an optical resonator [26,27]. To do so, we replace the single atom in the experimental setup in Figure 3 by a collection of N atoms. In analogy to Equation (9), the interaction Hamiltonian H_I between phonons and cavity photons now equals

$$H_I = \sum_{i=1}^{N} \hbar g_{\text{eff}}^{(i)} \left(b_i c^\dagger + b_i^\dagger c \right), \quad (14)$$

where $g_{\text{eff}}^{(i)}$ denotes the effective atom–cavity coupling constant of atom i. This coupling constant is essentially the same as g_{eff} in Equation (13) and depends in general on the position of atom i. Moreover, b_i denotes the phonon annihilation operator of atom i with $[b_i, b_j^\dagger] = \delta_{ij}$. To simplify the above Hamiltonian, we introduce a collective phonon annihilation operator B,

$$B = \frac{\sum_{i=1}^{N} g_{\text{eff}}^{(i)} b_i}{\tilde{g}_{\text{eff}}} \quad \text{with} \quad \tilde{g}_{\text{eff}} = \left(\sum_{i=1}^{N} |g_{\text{eff}}^{(i)}|^2 \right)^{1/2}, \quad (15)$$

with $[B, B^\dagger] = 1$. Using this notation, H_I in Equation (14) simplifies to

$$H_I = \hbar \tilde{g}_{\text{eff}} \left(Bc^\dagger + B^\dagger c \right). \quad (16)$$

Notice that the effective coupling constant $\tilde{g}_{\text{eff}}$ scales as the square root of the number of atoms N inside the cavity. For example, if all atomic particles couple equally to the cavity field with a coupling constant $g_{\text{eff}} \equiv g_{\text{eff}}^{(i)}$, then $\tilde{g}_{\text{eff}} = \sqrt{N} g_{\text{eff}}$. This means, in the case of many atoms, the effective phonon–photon coupling is collectively enhanced [26].

When comparing H_I in Equation (9) with H_I in Equation (14), we see that both Hamiltonians are essentially the same. Moreover, the density matrix ρ_I obeys the master equation in Equation (10) in both cases. Hence, we expect the same cooling dynamics in the one atom and in the many atom case.

Suppose all atoms experience the same atom–cavity coupling constant g_{eff}, the effective cooling rate of the common vibrational mode B becomes

$$\gamma_{N\,\text{atoms}} = N g_{\text{eff}}^2 / \kappa, \tag{17}$$

in analogy to Equation (13). This cooling rate is N times larger than the cooling rate which we predicted in the previous subsection for cavity-mediated laser cooling of a single atom. Using sufficiently large number of atoms N, it is therefore possible to realize cooling rates $\gamma_{N\,\text{atoms}}$ with

$$\gamma_{N\,\text{atoms}} \gg \gamma_{1\,\text{atom}}^{\text{standard}}. \tag{18}$$

This suggests that the cooling rate of cavity-mediated laser cooling, i.e., the rate of change of the mean number n of B phonons in the system, is comparable and might even exceed the cooling rates of standard laser cooling of single trapped ions.

However, the above discussion also shows that cavity-mediated collective laser cooling only removes phonons from a single common vibrational mode B, while all other vibrational modes of the atomic gas do not experience the cooling laser. Once the B mode reaches its stationary state, the conversion of thermal energy into light stops. To nevertheless take advantage of the relatively high cooling rates of cavity-mediated collective laser cooling, an additional mechanism is needed [27,28]. As we shall see in the next section, one way of transferring energy between different vibrational modes is to intersperse cooling stages with thermalization stages (cf. Figure 4). The purpose of the cooling stages is to rapidly remove energy from the system. The purpose of subsequent thermalization stages is to transfer energy from the surroundings of the bubble and from the different vibrational modes of the atoms into the B mode. Repeating thermalization and cooling stages is expected to result in the cooling of the whole setup in Figure 1.

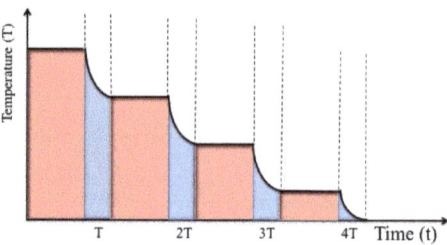

Figure 4. Schematic view of the expected dynamics of the temperature of the atomic gas during cavity-mediated collective laser cooling which involves a sequence of cooling stages (blue) and thermalization stages (pink). During thermalization stages, heat is transferred from the different vibrational degrees of freedoms of the atoms into a certain collective vibrational mode B, while the mean temperature of the atoms remains the same. During cooling stages, energy from the B mode into light. Eventually, the atomic gas becomes very cold.

3. Thermalization of an Atomic Gas with Elastic Collisions

Thermalization stages occur naturally in cavitating bubbles between collapse stages due to elastic collisions. As we show below, these transfer an atomic gas into its thermal state, thereby re-distributing energy between all if its vibrational degrees of freedom. During bubble expansions, the phonon frequencies of the atoms become very small. It is therefore safe to assume that the atoms do not see the cooling laser during thermalization stages.

3.1. The Thermal State of a Single Harmonic Oscillator

As in the previous section, we first consider a single trapped atom inside a harmonic trapping potential. Its thermal state equals [47]

$$\rho = \frac{1}{Z} e^{-\beta H} \quad \text{with} \quad Z = \text{Tr}(e^{-\beta H}), \tag{19}$$

where H is the relevant harmonic oscillator Hamiltonian, $\beta = 1/k_B T$ is the thermal Lagrange parameter for a given temperature T, k_B is Boltzmann's constant and Z denotes the partition function which normalizes the density matrix ρ of the atom. For sufficiently large atomic transition frequencies ω_0, the thermal state of the atom is to a very good approximation given by its ground state $|g\rangle$, unless the atom becomes very hot. In the following, we therefore neglect its electronic degrees of freedom. Hence, the Hamiltonian H in Equation (19) equals

$$H = \hbar \nu \left(b^\dagger b + \tfrac{1}{2} \right), \tag{20}$$

where ν and b denote again the frequency and the annihilation operator of a single phonon. Combining Equations (19) and (20), we find that [47]

$$Z = \frac{e^{-\frac{1}{2}\lambda}}{1 - e^{-\lambda}}. \tag{21}$$

with $\lambda = \beta \hbar \nu$. Here, we are especially interested in the expectation value of the thermal energy of the vibrational mode of the trapped atom which equals $\langle H \rangle = \text{Tr}(H\rho)$. Hence, using Equation (19), one can show that

$$\langle H \rangle = \frac{1}{Z} \text{Tr}\left(H e^{-\beta H}\right) = -\frac{1}{Z} \frac{\partial}{\partial \beta} Z = -\frac{\partial}{\partial \beta} \ln Z. \tag{22}$$

Finally, combining this result with Equation (21), we find that

$$\langle H \rangle = \hbar \nu \left(\frac{e^{-\lambda}}{e^{-\lambda} - 1} + \frac{1}{2} \right) \tag{23}$$

which is Planck's expression for the average energy of a single quantum harmonic oscillator. Moreover,

$$m = \frac{e^{-\lambda}}{e^{-\lambda} - 1}, \tag{24}$$

since the mean phonon number $m = \langle b^\dagger b \rangle$ relates to $\langle H \rangle$ via $m = \langle H \rangle / \hbar \nu - \tfrac{1}{2}$.

3.2. The Thermal State of Many Atoms with Collisions

Next we calculate the thermal state of a strongly confined atomic gas with strong elastic collisions. This situation has many similarities with the situation considered in the previous subsection. The atoms constantly collide with their respective neighbors which further increases the confinement of the individual particles. Hence, we assume in the following that the atoms no longer experience the phonon frequency ν but an increased phonon frequency ν_{eff}. If all atoms experience approximately the same interaction, their Hamiltonian H equals

$$H = \sum_{i=1}^{N} \hbar \left(\nu_{\text{eff}} + \tfrac{1}{2} \right) b_i^\dagger b_i \tag{25}$$

to a very good approximation. Here, b_i denotes again the phonon annihilation operator of atom i. Comparing this Hamiltonian with the harmonic oscillator Hamiltonian in Equation (20) and substituting H in Equation (25) into Equation (19) to obtain the thermal state of many atoms, we find that this thermal state is simply the product of the thermal states of the individual atoms. All atoms have the same thermal state, their mean phonon number $m_i = \langle b_i^\dagger b_i \rangle$ equals

$$m_i = \frac{e^{-\lambda_{\text{eff}}}}{e^{-\lambda_{\text{eff}}} - 1} \qquad (26)$$

with $\lambda_{\text{eff}} = \hbar \nu_{\text{eff}}/k_B T$, in analogy to Equation (24). This equation shows that any previously depleted collective vibrational mode of the atoms becomes re-populated during thermalization stages.

4. A Quantum Heat Exchanger with Cavitating Bubbles

As pointed out in Section 1, the aim of this paper is to design a quantum heat exchanger for nanotechnology. The proposed experimental setup consists of a liquid on top of the device which we aim to keep cool, a transducer and a cooling laser (cf. Figure 1). The transducer generates cavitating bubbles which need to contain atomic particles and whose diameters need to change very rapidly in time. The purpose of the cooling laser is to stimulate the conversion of heat into light. The cooling of the atomic particles inside cavitating bubbles subsequently aids the cooling of the liquid which surrounds the bubbles and its environment via adiabatic heat transfers.

To gain a better understanding of the experimental setup in Figure 1, Section 4.1 describes the main characteristics of single bubble sonoluminescence experiments [32–36]. Section 4.2 emphasizes that there are many similarities between sonoluminescence and quantum optics experiments [29,30]. From this, we conclude that sonoluminescence experiments naturally provide the main ingredients for the implementation of cavity-mediated collective laser cooling of an atomic gas [26–28]. Finally, in Sections 4.3 and 4.4, we have a closer look at the physics of the proposed quantum heat exchanger and estimate cooling rates.

4.1. Single Bubble Sonoluminescence Experiments

Sonoluminescence can be defined as a phenomenon of strong light emission from collapsing bubbles in a liquid, such as water [32–34]. These bubbles need to be filled with noble gas atoms, such as nitrogen atoms, which occur naturally in air. Alternatively, the bubbles can be filled with ions from ionic liquids, molten salts, and concentrated electrolyte solutions [48]. Moreover, the bubbles need to be acoustically confined and periodically driven by ultrasonic frequencies. As a result, the bubble radius changes periodically in time, as illustrated in Figure 5. The oscillation of the bubble radius regenerates itself with unusual precision.

At the beginning of every expansion phase, the bubble oscillates about its equilibrium radius until it returns to its fastness. During this process, the bubble temperature changes adiabatically and there is an exchange of thermal energy between the atoms inside the bubble and the surrounding liquid. During the collapse phase of a typical single-bubble sonoluminescence, i.e., when the bubble reaches its minimum radius, its inside becomes thermally isolated from the surrounding environment and the atomic gas inside the bubble becomes strongly confined. Usually, a strong light flash occurs at this point which is accompanied by a sharp increase of the temperature of the particles. Experiments have shown that increasing the concentration of atoms inside the bubble increases the intensity of the emitted light [35,36].

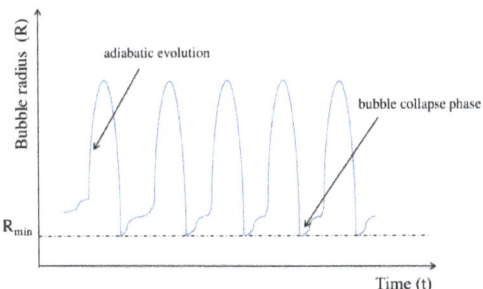

Figure 5. Schematic view of the time dependence of the bubble radius in a typical single-bubble sonoluminesence experiment. Most of the time, the bubble evolves adiabatically and exchanges thermal energy with its surroundings. However, at regular time intervals, the bubble radius suddenly collapses. At this point, the bubble becomes thermally isolated. When it reaches its minimum radius, the system usually emits a strong flash of light in the optical regime.

4.2. A Quantum Optics Perspective on Sonoluminescence

The above observations suggest many similarities between sonoluminescence and quantum optics experiments with trapped atomic particles [29,30]. When the bubble reaches its minimum radius, an atomic gas becomes very strongly confined [31]. The quantum character of the atomic motion can no longer be neglected and, as in ion trap experiments (cf. Section 2.1), the presence of phonons with different trapping frequencies ν has to be taken into account. Moreover, when the bubble reaches its minimum radius, its surface can become opaque and almost metallic [37]. When this happens, the bubble traps light inside and closely resembles an optical cavity which can be characterized by a frequency $\omega_{\rm cav}$ and a spontaneous decay rate κ. Since the confined particles have atomic dipole moments, they naturally couple to the quantized electromagnetic field inside the cavity. The result can be an exchange of energy between atomic dipoles and the cavity mode. The creation of photons inside the cavity is always accompanied by a change of the vibrational states of the atoms. Hence, the subsequent spontaneous emission of light in the optical regime results in a permanent change of the temperature of the atomic particles.

A main difference between sonoluminescence and cavity-mediated collective laser cooling is the absence and presence of external laser driving (cf. Section 2.3). However, even in the absence of external laser driving, there can be a non-negligible amount of population in the excited atomic states $|e\rangle$. This applies, for example, if the atomic gas inside the cavitating bubble is initially prepared in the thermal equilibrium state of a finite temperature T. Once surrounded by an optical cavity, as it occurs during bubble collapse phases, excited atoms can return into their ground state via the creation of a cavity photon (cf. Figure 6). Suddenly, an additional de-excitation channel has become available to them. As pointed out in Refs. [29,30], the creation a cavity photons is more likely accompanied by the creation of a phonon than the annihilation of a phonon since

$$\begin{aligned} B^\dagger &= \sum_{m=0}^{\infty} \sqrt{m+1}\, |m+1\rangle\langle m|, \\ B &= \sum_{m=0}^{\infty} \sqrt{m}\, |m-1\rangle\langle m|. \end{aligned} \qquad (27)$$

Here, B and $B^\dagger$ denote the relevant phonon annihilation and creation operators, while $|m\rangle$ denotes a state with exactly m phonons. As one can see from Equation (27), the normalization factor of $B^\dagger |m\rangle$ is slightly larger than the normalization factor of the state $B |m\rangle$. When the cavity photon is subsequently lost via spontaneous photon emission, the newly-created phonon remains inside the bubble. Hence, the light emission during bubble collapse phases is usually accompanied by heating, until the sonoluminescing bubble reaches an equilibrium.

During each bubble collapse phase, cavitating bubbles are thermally isolated from their surroundings. However, during the subsequent expansion phase, system parameters change adiabatically and there is a constant exchange of thermal energy between atomic gas inside the bubble and the surrounding liquid (cf. Figure 5). Eventually, the atoms reach an equilibrium between heating during bubble collapse phases and the loss of energy during subsequent expansion phases. Experiments have shown that the atomic gas in side the cavitating bubble can reaches temperature of the order of 10^4 K which strongly supports the hypothesis that there is a very strong coupling between the vibrational and the electronic states of the confined particles [35,36].

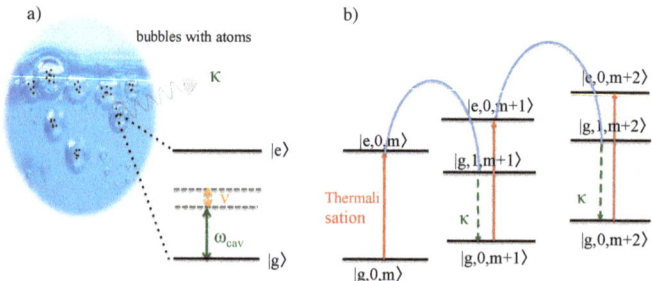

Figure 6. (a) From a quantum optics point of view, one of the main characteristics of sonoluminescence experiments is that cavitating bubbles provide a very strong confinement for atomic particles. This means that the quantum character of their motional degrees of freedom has to be taken into account. As in ion trap experiments, we denote the corresponding phonon frequency in this paper by ν. Moreover, during its collapse phase, the surface of the bubble becomes opaque and confines light, thereby forming an optical cavity with frequency ω_{cav} and a spontaneous decay rate κ. (b) Even in the absence of external laser driving, some of the atoms are initially in their excited state $|e\rangle$ due to being prepared in a thermal equilibrium state at a finite temperature T. When returning into their ground state via the creation of a cavity photon, which is only possible during the bubble collapse phase, most likely a phonon is created. This creation of phonons implies heating. Indeed, sonoluminescence experiments often reach relatively high temperatures [35,36].

4.3. Cavity-Mediated Collective Laser Cooling of Cavitating Bubbles

The previous subsection shows that, during each collapse phase, the dynamics of the cavitating bubbles in Figure 1 is essentially the same as the dynamics of the experimental setup in Figure 3 but with the single atom replaced by an atomic gas. When the bubble reaches its minimum diameter d_{min}, it forms an optical cavity which supports a discrete set of frequencies ω_{cav},

$$\omega_{cav} = j \times \frac{\pi c}{d_{min}}, \tag{28}$$

where c denotes the speed of light in air and $j = 1, 2, \ldots$ is an integer. As illustrated in Figure 7, the case $j = 1$ corresponds to a cavity photon wavelength $\lambda_{cav} = 2d_{min}$. Moreover, $j = 2$ corresponds to $\lambda_{cav} = d_{min}$, and so on. Under realistic conditions, the cavitating bubbles are not all of the same size which is why every j is usually associated with a range of frequencies ω_{cav} (cf. Figure 7). Here, we are especially interested in the parameter j, where the relevant cavity frequencies lie in the optical regime. All other parameters j can be neglected, once a laser field with an optical frequency ω_L is applied, if neighboring frequency bands are sufficiently detuned.

In addition, we know that the phonon frequency ν of the collective phonon mode B assumes its maximum ν_{max} during the bubble collapse phase. Suppose the cavity detuning $\Delta_{cav} = \omega_L - \omega_{cav}$ of the applied laser field is chosen such that

$$\Delta_{cav} \sim \nu_{max} \quad \text{and} \quad \nu_{max} \geq \kappa, \tag{29}$$

in analogy to Equation (2). As we have seen in Section 2.3, in this case, the two-step transition which results in the simultaneous annihilation of a phonon and the creation of a cavity photon becomes resonant and dominates the system dynamics. If the creation of a cavity photon is followed by a spontaneous emission, the previously annihilated phonon cannot be restored and is permanently lost. Overall, we expect this cooling process to be very efficient, since the atoms are strongly confined and cavity cooling rates are collectively enhanced (cf. Equation (17)).

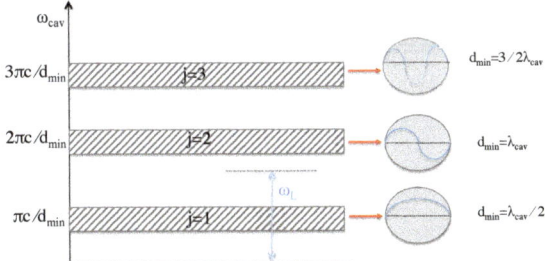

Figure 7. When the cavitating bubbles inside the liquid reach their minimum diameters d_{min}, their walls become opaque and trap light on the inside. To a very good approximation, they form cavities which can be described by spontaneous decay rates κ and cavity frequencies ω_{cav} (cf. Equation (28)). Suppose the diameters of the bubbles inside the liquid occupy a relatively small range of values. Then, every integer number j in Equation (28) corresponds to a relatively narrow range of cavity frequencies ω_{cav}. Here, we are especially interested in the parameter j for which the ω_{cav}'s lie in the optical regime. When this applies, we can apply a cooling laser with an optical frequency ω_L which can cool the atoms in all bubbles. Some bubbles will be cooled more efficiently than others. However, as long as the relevant frequency bands are relatively narrow, none of the bubbles will be heated.

To cool not only very tiny but larger volumes, the experimental setup in Figure 1 should contain a relatively large number of cavitating bubbles. Depending on the quality of the applied transducer, the minimum diameters d_{min} of these bubbles might vary in size. Consequently, the collection of bubbles supports a finite range of cavity frequencies ω_{cav} (cf. Figure 7 so that it becomes impossible to realize the ideal cooling condition $\Delta_{cav} \sim \nu_{max}$ in Equation (29) for all bubbles. However, as long as the frequency ω_L of the cooling laser is smaller than all optical cavity frequencies ω_{cav}, the system dynamics will be dominated by cooling and not by heating. In general, it is important that the diameters of the bubbles does not vary by too much.

Section 2.3 also shows that cavity-mediated collective laser cooling only removes thermal energy from a single collective vibrational mode B of the atoms. Once this mode is depleted, the cooling process stops. To efficiently cool an entire atomic gas, a mechanism is needed which rapidly re-distributes energy between different vibrational degrees of freedom, for example, via thermalization based on elastic collisions (cf. Section 3). As shown above, between cooling stages, cavitating bubbles evolve essentially adiabatically and the atoms experience strong collisions. In other words, the expansion phase of cavitating bubbles automatically implements the intermittent thermalization stages of cavity-mediated collective laser cooling.

Finally, let us point out that it does not matter whether the cooling laser is turned on or off during thermalization stages, i.e., during bubble expansion phases. As long as optical cavities only form during the bubble collapse phases, the above-described conversion of heat into light only happens when the bubble reaches its minimum diameter. The reason for this is that noble gas atoms, such as nitrogen, have very large transition frequencies ω_0. The direct laser excitation of atomic particles is therefore relatively unlikely, even when the cooling laser is turned on. If we could excite the atoms directly by laser driving, we could cool them even more efficiently (cf. Section 2.1).

4.4. Cooling of the Surroundings via Heat Transfer

The purpose of the heat exchanger which we propose here is to constantly remove thermal energy from the liquid surrounding the cavitating bubbles and device on which the liquid is placed (cf. Figure 1). As described in the previous subsection, the atomic gas inside the bubbles is cooled by very rapidly converting heat into light during each collapse phase. In between collapse phases, the cavitating bubbles evolve adiabatically and naturally cool their immediate environment via heat transfer. As illustrated in Figure 8, alternating cooling and thermalization stages (or collapse and expansion phases) is expected to implement a quantum heat exchanger, which does not require the actual transport of particles from one place to another.

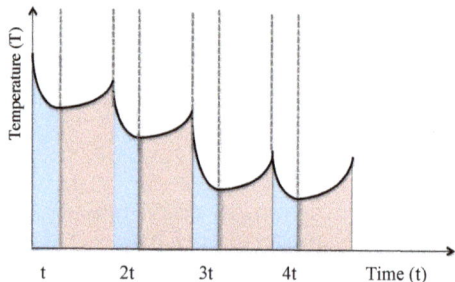

Figure 8. Schematic view of the expected dynamics of the temperature of a confined atomic gas during bubble collapse stages (blue) and expansion stages (pink). During expansion stages, heat is transferred from the outside into the inside of the bubble, thereby increasing the temperature of the atoms. During bubble collapse stages, heat is converted into light, thereby resulting in the cooling of the system in Figure 1. Eventually, both processes balance each other out and the temperature of the system remains constant on a coarse grained time scale.

Finally, let us have a closer look at achievable cooling rates for micro- and nanotechnology devices with length dimensions in the nano- and micrometer regime. Unfortunately, we do not know how rapidly heat can be transferred from the nanotechnology device to the liquid and from there to the atomic gas inside the cavitating bubbles. However, any thermal energy which is taken from the atoms comes eventually from the environment which we aim to cool. Suppose the relevant phonon frequencies ν_{max} are sufficiently large to ensure that every emitted photon indicates the loss of one phonon, i.e., the loss of one energy quantum $\hbar\nu_{max}$. Moreover, suppose our quantum heat exchanger contains a certain amount of liquid, let us say water, of mass m_{water} and heat capacity $c_{water}(T)$ at an initial temperature T_0. Then, we can ask the question: How many photons $N_{photons}$ do we need to create in order to cool the water by a certain temperature ΔT?

From thermodynamics, we know that the change in the thermal energy of the water equals

$$\Delta Q = c_{water}(T_0) m_{water} \Delta T \qquad (30)$$

in this case. Moreover, we know that

$$\Delta Q = N_{photons} \hbar \nu_{max}. \qquad (31)$$

Hence, the number of photons that needs to be produced is given by

$$N_{photons} = \frac{c_{water}(T_0) \, m_{water} \, \Delta T}{\hbar \nu_{max}}. \qquad (32)$$

The time t_{cool} it would take to create this number of photons equals

$$t_{cool} = \frac{N_{photons}}{N_{atoms} \, I}, \qquad (33)$$

where I denotes the average single-atom photon emission rate and N_{atoms} is the number of atoms involved in the cooling process. When combining the above equations, we find that the cooling rate $\gamma_{cool} = t_{cool}/\Delta T$ of the proposed cooling process equals

$$\gamma_{cool} = \frac{c_{water}(T) \, m_{water}}{N_{atoms} \, I \, \hbar \nu_{max}} \qquad (34)$$

to a very good approximations.

As an example, suppose we want to cool one cubic micrometer of water ($V_{water} = 1 \, \mu m^3$) at room temperature ($T_0 = 20\,°C$). In this case, $m_{water} = 10^{-15}$ g and $c_{water}(T_0) = 4.18$ J/gK to a very good approximation. Suppose $\nu = 100$ MHz (a typical frequency in ion trap experiments is $\nu = 10$ MHz), $I = 10^6/$s and $N_{atoms} = 10^8$ (a typical bubble in single bubble sonoluminescence contains about 10^8 atoms). Substituting these numbers into Equation (33) yields a cooling rate of

$$\gamma_{cool} = 3.81 \, \text{ms/K}. \qquad (35)$$

Achieving cooling rates of the order of Kelvin temperatures per millisecond seems therefore experimentally feasible. As one can see from Equation (33), to reduce cooling rates further, one can either reduce the volume that requires cooling, increase the number of atoms involved in the cooling process or increase the trapping frequency ν_{max} of the atomic gas inside collapsing bubbles. All of this is, at least in principle, possible.

5. Conclusions

In this paper, we point out similarities between quantum optics experiments with strongly confined atomic particles and single bubble sonoluminescence experiments [29,30]. In both situations, interactions are present, which can be used to convert thermal energy very efficiently into light. When applying an external cooling laser to cavitating bubbles, as illustrated in Figure 1, we therefore expect a rapid transfer of heat into light which can eventually result in the cooling of relatively small devices. Our estimates show that it might be possible to achieve cooling rates of the order of milliseconds per Kelvin temperatures for cubic micrometers of water. The proposed quantum heat exchanger is expected to find applications in research experiments and in micro- and nanotechnology. A closely related cooling technique, namely laser cooling of individually trapped ions, already has a wide range of applications in quantum technology [9–16].

Author Contributions: A.A. and A.B. are quantum opticians and contributed the design of the proposed quantum heat exchanger and its formal analysis. S.A.P. helped to improve the initial design and provided additional insight into micro- and nanoscale physics with cavitating bubbles. All authors have read and agreed to the published version of the manuscript.

Funding: A.A. acknowledges funding from The Ministry of Education in Saudi Arabia. In addition, this work was supported by the United Kingdom Engineering and Physical Sciences Research Council (EPSRC) from the Oxford Quantum Technology Hub NQIT (grant number EP/M013243/1). Statement of compliance with EPSRC policy framework on research data: This publication is theoretical work that does not require supporting research data.

Conflicts of Interest: The authors declare no conflict of interest.

References

1. Hänsch, T.W.; Schawlow, A.L. Cooling of gases by laser radiation. *Opt. Commun.* **1975**, *13*, 68–69. [CrossRef]
2. Wineland, D.J.; Dehmelt, H. Proposed $10^{14}\,\Delta\nu < \nu$ laser fluorescence spectroscopy on Tl$^+$ mono-ion oscillator III. *Bull. Am. Phys. Soc.* **1975**, *20*, 637.
3. Wineland, D.; Drullinger, R.; Walls, F. Radiation-pressure cooling of bound-resonant absorbers. *Phys. Rev. Lett.* **1978**, *40*, 1639–1642. [CrossRef]
4. Neuhauser, W.; Hohenstatt, M.; Toschek, P.; Dehmelt, H. Optical-sideband cooling of visible atom cloud confined in parabolic well. *Phys. Rev. Lett.* **1978**, *41*, 233–236. [CrossRef]
5. Leibfried, D.; Blatt, R.; Monroe, C.; Wineland, D. Quantum dynamics of single trapped ions. *Rev. Mod. Phys.* **2003**, *75*, 281–324. [CrossRef]
6. Chu, S. Nobel Lecture: The manipulation of neutral particles. *Rev. Mod. Phys.* **1998**, *70*, 685–706. [CrossRef]
7. Phillips, W.D. Laser cooling and trapping of neutral atoms. *Rev. Mod. Phys.* **1998**, *70*, 721–741. [CrossRef]
8. Rosenband, T.; Hume, D.B.; Schmidt, P.O.; Chou, C.W.; Brusch, A.; Lorini, L.; Oskay, W.H.; Drullinger, R.E.; Fortier, T.M.; Stalnaker, J.E.; et al. Frequency Ratio of Al$^+$ and Hg$^+$ Single-Ion Optical Clocks; Metrology at the 17th Decimal Place. *Science* **2008**, *319*, 1808–1812. [CrossRef]
9. Ludlow, A.D.; Boyd, M.M.; Ye, J.; Peik, E.; Schmidt, P.O. Optical atomic clocks. *Rev. Mod. Phys.* **2015**, *87*, 637–701. [CrossRef]
10. Schmidt-Kaler, F.; Häffner, H.; Riebe, M.; Gulde, S.; Lancaster, G.P.T.; Deuschle, T.; Becher, C.; Roos, C.F.; Eschner, J.; Blatt, R. Realization of the Cirac-Zoller controlled-NOT quantum gate. *Nature* **2003**, *422*, 408–411. [CrossRef]
11. Leibfried, D.; DeMarco, B.; Meyer, V.; Lucas, D.; Barrett, M.; Britton, J.; Itano, W.M.; Jelenkovic, B.; Langer, C.; Rosenband, T.; et al. Experimental demonstration of a robust, high-fidelity geometric two ion-qubit phase gate. *Nature* **2003**, *422*, 412–415. [CrossRef] [PubMed]
12. Debnath, S.; Linke, N.M.; Figgatt, C.; Landsman, K.A.; Wright, K.; Monroe, C. Demonstration of a small programmable quantum computer with atomic qubits. *Nature* **2016**, *536*, 63–66. [CrossRef]
13. Stephenson, L.J.; Nadlinger, D.P.; Nichol, B.C.; An, S.; Drmota, P.; Ballance, T.G.; Thirumalai, K.; Goodwin, J.F.; Lucas, D.M.; Ballance, C.J. High-rate, high-fidelity entanglement of qubits across an elementary quantum network. *arXiv* **2019**, arXiv:1911.10841.
14. Porras, D.; Cirac, J.I. Effective Quantum Spin Systems with Trapped Ions. *Phys. Rev. Lett.* **2004**, *92*, 207901. [CrossRef] [PubMed]
15. Barreiro, J.T.; Müller, M.; Schindler, P.; Nigg, D.; Monz, T.; Chwalla, M.; Hennrich, M.; Roos, C.F.; Zoller, P.; Blatt, R. An open-system quantum simulator with trapped ions. *Nature* **2011**, *470*, 486–491. [CrossRef]
16. Maiwald, R.; Leibfried, D.; Britton, J.; Bergquist, J.C.; Leuchs, G.; Wineland, D.J. Stylus ion trap for enhanced access and sensing. *Nat. Phys.* **2009**, *5*, 551–554. [CrossRef]
17. Stick, D.; Hensinger, W.K.; Olmschenk, S.; Madsen, M.J.; Schwab, K.; Monroe, C. Ion trap in a semiconductor chip. *Nat. Phys.* **2006**, *2*, 36–39. [CrossRef]
18. Goodwin, J.F.; Stutter, G.; Thompson, R.C.; Segal, D.M. Resolved-Sideband Laser Cooling in a Penning Trap. *Phys. Rev. Lett.* **2016**, *116*, 143002. [CrossRef]
19. Stenholm, S. The semiclassical theory of laser cooling. *Rev. Mod. Phys.* **1986**, *58*, 699–739. [CrossRef]
20. Blake, T.; Kurcz, A.; Saleem, N.S.; Beige, A. Laser cooling of a trapped particle with increased Rabi frequencies. *Phys. Rev. A* **2011**, *84*, 053416. [CrossRef]
21. Hsu, T.R. *MEMS & Microsystems: Design and Manufacture*; McGraw-Hill: New York, NY, USA, 2002.
22. Kim, P.; Shi, L.; Majumdar, A.; McEuen, P.L. Thermal Transport Measurements of Individual Multiwalled Nanotubes. *Phys. Rev. Lett.* **2001**, *87*, 215502. [CrossRef] [PubMed]
23. Saniei, N. Nanotechnology and heat transfer. *Heat Transf. Eng.* **2007**, *28*, 255–257. [CrossRef]
24. Domokos, P.; Ritsch, H. Collective Cooling and Self-Organization of Atoms in a Cavity. *Phys. Rev. Lett.* **2002**, *89*, 253003. [CrossRef] [PubMed]
25. Ritsch, H.; Domokos, P.; Brennecke, F.; Esslinger, T. Cold atoms in cavity-generated dynamical optical potentials. *Rev. Mod. Phys.* **2013**, *85*, 553–601. [CrossRef]
26. Beige, A.; Knight, P.L.; Vitiello, G. Cooling many particles at once. *New J. Phys.* **2005**, *7*, 96. [CrossRef]
27. Kim, O.; Deb, P.; Beige, A. Cavity-mediated collective laser-cooling of a non-interacting atomic gas inside an asymmetric trap to very low temperatures. *J. Mod. Opt.* **2018**, *65*, 693–705. [CrossRef]

28. Aljaloud, A.; Beige, A. Cavity-mediated collective laser cooling of atoms inside cavitating bubbles. 2020, in preparation.
29. Kurcz, A.; Capolupo, A.; Beige, A. Sonoluminescence and quantum optical heating. *New J. Phys.* **2009**, *11*, 053001. [CrossRef]
30. Beige, A.; Kim, O. A cavity-mediated collective quantum effect in sonoluminescing bubbles. *J. Phys. Conf. Ser.* **2015**, *656*, 012177. [CrossRef]
31. Moss, C.W. Understanding the periodic driving pressure in the Rayleigh-Plesset equation. *J. Acoust. Soc. Am.* **1997**, *101*, 1187–1190. [CrossRef]
32. Gaitan, D.F.; Crum, L.A.; Church, C.C.; Roy, R.A. Sonoluminescence and bubble dynamics for a single, stable, cavitation bubble. *J. Acoust. Soc. Am.* **1992**, *91*, 3166–3183. [CrossRef]
33. Benner, M.P.; Hilgenfeldt, S.; Lohse, D. Single-bubble sonoluminescence. *Rev. Mod. Phys.* **2002**, *74*, 425–484. [CrossRef]
34. Camara, C.; Putterman, S.; Kirilov, E. Sonoluminescence from a single bubble driven at 1 megahertz. *Phys. Rev. Lett.* **2004**, *92*, 124301. [CrossRef]
35. Flannigan, D.J.; Suslick, K.S. Plasma formation and temperature measurement during single-bubble cavitation. *Nature* **2005**, *434*, 52–55. [CrossRef]
36. Suslick, K.S.; Flannigan, D.J. Inside a collapsing bubble: Sonoluminescence and the conditions during cavitation. *Annu. Rev. Phys. Chem.* **2008**, *59*, 659–683. [CrossRef] [PubMed]
37. Khalid, S.; Kappus, B.; Weninger, K.; Putterman, S. Opacity and transport measurements reveal that dilute plasma models of sonoluminescence are not valid. *Phys. Rev. Lett.* **2012**, *108*, 104302. [CrossRef] [PubMed]
38. Daul, J.-M.; Grangier, P. Cavity-induced damping and level shifts in a wide aperture spherical resonator. *Eur. Phys. J. D* **2005**, *32*, 181–194. [CrossRef]
39. Daul, J.-M.; Grangier, P. Vacuum field atom trapping in a wide aperture spherical resonator. *Eur. Phys. J. D* **2005**, *32*, 195–200. [CrossRef]
40. Cao, G.; Danworaphong, S.; Diebold, G.J. A search for laser heating of a sonoluminescing bubble. *Eur. Phys. J. Spec. Top.* **2008**, *153*, 215–221. [CrossRef]
41. Suslick, K.S. Sonochemistry. *Science* **1990**, *247*, 1439–1445. [CrossRef]
42. Cirac, J.I.; Parkins, A.S.; Blatt, R.; Zoller, P. Cooling of a trapped ion coupled strongly to a quantized cavity mode. *Opt. Commun.* **1993**, *97*, 353–359. [CrossRef]
43. Cirac, J.I.; Lewenstein, M.; Zoller, P. Laser cooling a trapped atom in a cavity: Bad-cavity limit. *Phys. Rev. A* **1995**, *51*, 1650–1655. [CrossRef] [PubMed]
44. Blake, T.; Kurcz, A.; Beige, A. Comparing cavity and ordinary laser cooling within the Lamb-Dicke regime. *J. Mod. Opt.* **2011**, *58*, 1317–1328. [CrossRef]
45. Kim, O.; Beige, A. Mollow triplet for cavity-mediated laser cooling. *Phys. Rev. A* **2013**, *88*, 053417. [CrossRef]
46. Urunuela, E.; Alt, W.; Keiler, E.; Meschede, D.; Pandey, D.; Pfeifer, H.; Macha, T. Ground-state cooling of a single atom inside a high-bandwidth cavity. *Phys. Rev. A* **2020**, *101*, 023415. [CrossRef]
47. Blaise, P.; Henri-Rousseau, O. *Quantum Oscillators*; John Wiley and Sons: New York, NY, USA, 2011.
48. Flannigan, D.J.; Hopkins, S.D.; Suslick, K.S. Sonochemistry and sonoluminescence in ionic liquids, molten salts, and concentrated electrolyte solutions. *J. Organomet. Chem.* **2005**, *690*, 3513–3517. [CrossRef]

© 2020 by the authors. Licensee MDPI, Basel, Switzerland. This article is an open access article distributed under the terms and conditions of the Creative Commons Attribution (CC BY) license (http://creativecommons.org/licenses/by/4.0/).

Article

Quantifying Athermality and Quantum Induced Deviations from Classical Fluctuation Relations

Zoë Holmes *,†, Erick Hinds Mingo *,†, Calvin Y.-R. Chen and Florian Mintert

Controlled Quantum Dynamics Theory Group, Imperial College London, London SW7 2BW, UK; calvin.chen16@imperial.ac.uk (C.Y.-R.C.); f.mintert@imperial.ac.uk (F.M.)
* Correspondence: z.holmes15@ic.ac.uk (Z.H.); e.hinds-mingo15@imperial.ac.uk (E.H.M.)
† These authors contributed equally to this work.

Received: 20 December 2019; Accepted: 14 January 2020; Published: 16 January 2020

Abstract: In recent years, a quantum information theoretic framework has emerged for incorporating non-classical phenomena into fluctuation relations. Here, we elucidate this framework by exploring deviations from classical fluctuation relations resulting from the athermality of the initial thermal system and quantum coherence of the system's energy supply. In particular, we develop Crooks-like equalities for an oscillator system which is prepared either in photon added or photon subtracted thermal states and derive a Jarzynski-like equality for average work extraction. We use these equalities to discuss the extent to which adding or subtracting a photon increases the informational content of a state, thereby amplifying the suppression of free energy increasing process. We go on to derive a Crooks-like equality for an energy supply that is prepared in a pure binomial state, leading to a non-trivial contribution from energy and coherence on the resultant irreversibility. We show how the binomial state equality fits in relation to a previously derived coherent state equality and offers a richer feature-set.

Keywords: fluctuation relation; Crooks equality; quantum thermodynamics; coherence; athermality; photon added thermal state; photon subtracted thermal state; binomial states; generalised coherent states

1. Introduction

Thermodynamics, a theory of macroscopic systems at equilibrium, is vastly successful with a diverse range of applications [1–6]. This is perhaps somewhat surprising given the prevalence of non-equilibrium states and processes in nature. Underpinning this success is the second law of thermodynamics, an inequality that holds for all equilibrium and non-equilibrium processes alike [7]. However, the implication of an irreversible flow in the dynamics belies the "arrow of time", since the underlying laws of motion generally define no preferred temporal order [8]. A resolution to this seeming discrepancy arose in the form of fluctuation theorems, which derive the irreversibility beginning from time-reversal invariant dynamics [8–12].

The challenge of generalising fluctuation relations to quantum systems has attracted significant attention in recent years. The simplest approach defines the work done on a closed system as the change in energy found by performing projective measurements on the system at the start and end of the non-equilibrium process [10,13–17]. Extensions to this simple protocol have focused on formulations in terms of quantum channels [18–20], generalisations to open quantum systems [21,22] and alternative definitions for quantum work including those using quasi-probabilities [23,24], the consistent histories framework [25] and the quantum jump approach [26–28]. However, these approaches tend to be limited to varying degrees by the unavoidable impact of measurements on quantum systems. By defining quantum work in terms of a pair of projective measurements or continual weak measurements, the role of coherence is attenuated.

A new framework for deriving quantum fluctuation relations has recently emerged [29–32] which aims to fully incorporate non-classical thermodynamic effects into fluctuation relations by drawing on insights from the resource theory of quantum thermodynamics [33–39]. This framework considers an energy conserving and time reversal invariant interaction between an initially thermal system and a quantum *battery*, that is the energy source which supplies work to, or absorbs work from, the system. This framework can be taken as the starting point to derive Crooks-like relations for a harmonic oscillator battery prepared in coherent, squeezed and Schrödinger cat states [40]. These new equalities are used both to discuss coherence induced corrections to the Crooks equality and to propose an experiment to test the framework. Furthermore, the fluctuation relations give way to an interpretation involving coherent work states, a generalisation of Newtonian work for fully quantum dynamics. It was proved that the energetic and coherent properties of the coherent work is totally captured in this fluctuation setting [41].

In this paper, we use this new framework to explore deviations from classical fluctuation relations resulting from athermality of the initial thermal system and quantum coherence of the battery. In particular, we start by exploring the effects of athermality by developing Crooks equalities for a quantum harmonic oscillator system which is prepared in a photon added and photon subtracted thermal state. These states have received interest in quantum optics owing to their non-Gaussian and negative Wigner functions [42–44] along with their producibility in lab settings [42,45–47]. Furthermore, they have been suggested as useful resources in quantum key distribution [48], metrology [49] and continuous variable quantum computing [50,51], and there is growing interest in their thermodynamic properties [46,47].

We then proceed to investigate the role of coherence by deriving a Crooks equality for a battery prepared in pure binomial states. Binomial states can be viewed as analogues of coherent states for finite dimensional systems rather than infinite dimensional oscillators [52,53], leading to highly non-classical properties [54,55]. While binomial states are harder to produce in lab settings, there have been proposals [56,57]. The derived equality effectively generalises the coherent state Crooks equality of Holmes et al. [40], incorporating finite sized effects and leading to the coherent state equality in the appropriate limit. Moreover, binomial states quantify a smooth transition between semi-classical regimes and deep quantum regimes by encapsulating both coherent state and multi-qubit fluctuation relations in a single framework.

2. Background

2.1. Classical Fluctuation Relations

A system S is initially in thermal equilibrium with respect to Hamiltonian H_S^i at temperature T. It is then driven from equilibrium by a variation of Hamiltonian H_S^i to H_S^f, doing work W with probability $\mathcal{P}_F(W)$ in the process. This *forwards* process is compared to a *reverse* process in which a system thermalised with respect to H_S^f is pushed out of equilibrium by changing H_S^f to H_S^i, doing work $-W$ with probability $\mathcal{P}_R(-W)$. The ratio of these two probabilities is known as the Crooks equality [9],

$$\frac{\mathcal{P}_F(W)}{\mathcal{P}_R(-W)} = \exp\left(\beta(W - \Delta F)\right), \tag{1}$$

where ΔF is the equilibrium Helmholtz free energy difference and β is the inverse temperature $1/k_B T$.

The Crooks equality is a generalisation of the second law of thermodynamics. As a corollary to Crooks equality, one can derive the Jarzynski equality [12], which reads

$$\langle \exp(-\beta W) \rangle = \exp(-\beta \Delta F). \tag{2}$$

Finally, using Jensen's inequality [58], one arrives at the second law of thermodynamics in its formulation as a bound for the average extractable work $\langle W_{\text{ext}} \rangle \leq -\Delta F$. The Jarzynski equality has been used to calculate free energy changes for highly complex systems [59] such as unravelling of proteins [60], and as a theoretical tool to re-derive two of Einstein's key relations for Brownian motion and stimulated emission [61].

2.2. Fully Quantum Fluctuation Relations

Our starting point is a global "fully quantum fluctuation theorem" from [29], a more general relation than that explicated in [32,40,41], which can be used to derive a whole family of quantum fluctuation relations. A defining property of quantum systems is their ability to reside in superpositions of states belonging to different energy eigenspaces, a property often referred to simply as coherence. The quantum framework we present here carefully tracks the changes in these energetic coherences.

Changing the Hamiltonian of a system typically requires doing work or results in the system performing work and thus every fluctuation relation, at least implicitly, involves an energy source which supplies or absorbs this work. While often not explicitly modelled, the dynamics of the energy supply can contribute non-trivially to the evolution of the driven system. Thus, to enable a more careful analysis of the energy and coherence changes of the system, we consider an *inclusive* (this is in contrast to the *exclusionary* picture of the original Crooks and Jarzynski equalities) approach [29–31,40,41,62], which introduces a battery and assumes the system (S) and battery (B) evolve together under a time independent Hamiltonian H_{SB}.

To realise an effective change in system Hamiltonian from H_S^i to H_S^f with a time independent Hamiltonian, we assume a Hamiltonian of the form

$$H_{SB} = 1_S \otimes H_B + H_S^i \otimes \Pi_B^i + H_S^f \otimes \Pi_B^f \tag{3}$$

where H_B is the battery Hamiltonian and Π_B^i and Π_B^f are projectors onto two orthogonal subspaces, R_i and R_f, of the battery's Hilbert space. We assume the battery is initialised in a state in subspace R_i only and evolves under a unitary U to a final state in subspace R_f only, such that the system Hamiltonian is effectively time dependent, evolving from H_S^i to H_S^f.

To ensure that the energy supplied to the system is provided by the battery, we require the dynamics to be energy conserving such that $[U, H_{SB}] = 0$. We further assume that U and H_{SB} are time-reversal invariant with $U = \mathcal{T}(U)$ and $H_{SB} = \mathcal{T}(H_{SB})$. The time-reversal [63,64] operation $\mathcal{T}$ is defined as the transpose operation in the energy eigenbasis of the system and battery.

The most general process that can be described by a fluctuation relation within the inclusive framework involves preparing the system and battery in an initial state ρ, evolving it under the propagator U and then performing a measurement on the system and battery, which can be represented by the measurement operator X. The outcome of this measurement is quantified by

$$\mathcal{Q}(X|\rho) := \text{Tr}\left[X U \rho U^\dagger\right] \tag{4}$$

which can capture a number of different physical properties. For example, if the measurement operator X is chosen to be an observable, then $\mathcal{Q}(X|\rho)$ is the expectation value of the evolved state $U\rho U^\dagger$, whereas, if the measurement operator is chosen to be some state ρ', corresponding to the binary POVM measurement $\{\rho', 1-\rho'\}$, then $Q(\rho'|\rho)$ captures a transition probability between the state ρ and ρ' under the evolution U.

The global fluctuation relation relates $\mathcal{Q}(X_{SB}^f|\rho_{SB}^i)$ of a forwards process to $\mathcal{Q}(X_{SB}^i|\rho_{SB}^f)$ of a reverse process. For our purposes, we assume that the system and battery are initially uncorrelated in both the forwards and reverse processes, i.e.,

$$\rho_{SB}^i = \rho_S^i \otimes \rho_B^i \quad \text{and} \quad \rho_{SB}^f = \rho_S^f \otimes \rho_B^f \tag{5}$$

and suppose that independent measurements are made on the system and battery such that the measurement operator can be written in a separable form, i.e.,

$$X^i_{SB} = X^i_S \otimes X^i_B \quad \text{and} \quad X^f_{SB} = X^f_S \otimes X^f_B. \tag{6}$$

The global fluctuation relation holds for measurement operators and states related by the mapping $\mathcal{M}$ defined as

$$\rho^k_S = \mathcal{M}(X^k_S) \propto \mathcal{T}\left(\exp\left(-\frac{\beta H^k_S}{2}\right) X^k_S \exp\left(-\frac{\beta H^k_S}{2}\right)\right) \tag{7}$$

$$\rho^k_B = \mathcal{M}(X^k_B) \propto \mathcal{T}\left(\exp\left(-\frac{\beta H_B}{2}\right) X^k_B \exp\left(-\frac{\beta H_B}{2}\right)\right) \tag{8}$$

for $k = i, f$. This mapping arises naturally when one relates a forward and a reverse quantum process in the inclusive framework. When a measurement operator is a projection onto an energy eigenstate, then the state related by the mapping, Equation (7), is an energy eigenstate. Conversely, when no measurement is performed, i.e., $X = \mathbf{1}$, the corresponding state is a thermal state. However, in general, the mapping is non-trivial and essential to capture the influence of quantum coherence and athermality. The relationship between the four states quantified by the global fluctuation relation is sketched in Figure 1.

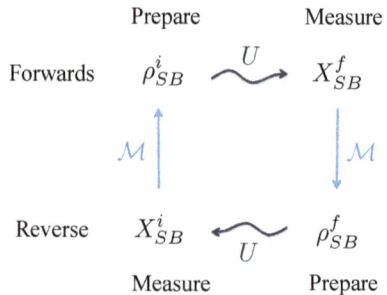

Figure 1. Relation between prepared states and measurements. In the forwards (reverse) process, the state $\rho^i_{SB} = \rho^i_S \otimes \rho^i_B$ ($\rho^f_{SB} = \rho^f_S \otimes \rho^f_B$) is prepared, it evolves under U as indicated by the wiggly arrow, and then the measurement $X^f_{SB} = X^f_S \otimes X^f_B$ ($X^i_{SB} = X^i_S \otimes X^i_B$) is performed. As indicated by the solid lines, the measurements X^i_{SB} and X^f_{SB} are related to the states ρ^i_{SB} and ρ^f_{SB}, respectively, by the mapping $\mathcal{M}$, defined in Equation (7).

For the uncorrelated initial states and measurement operators related by the mapping $\mathcal{M}$, the global fluctuation relation [29,40,41] can be written as

$$\frac{\mathcal{Q}(X^f_{SB}|\rho^i_{SB})}{\mathcal{Q}(X^i_{SB}|\rho^f_{SB})} = \exp\left(\beta(\Delta \tilde{W} - \Delta \tilde{F})\right) \tag{9}$$

in terms of the quantum generalisation

$$\Delta \tilde{F} := \tilde{E}(\beta, H^f_S, X^f_S) - \tilde{E}(\beta, H^i_S, X^i_S) \tag{10}$$

of the change in free energy, as well as a quantum generalisation of the work

$$\Delta \tilde{W} := \tilde{E}(\beta, H_B, X_B^i) - \tilde{E}(\beta, H_B, X_B^f). \tag{11}$$

supplied by the battery. The function

$$\tilde{E}(\beta, H, X) := -\frac{1}{\beta} \ln \left(\text{Tr} \left[\exp(-\beta H) X \right] \right) \tag{12}$$

is an *effective potential* that specifies the relevant energy value within the fluctuation theorem context. When the measurement operator is equal to the identity operation the effective potential, $\tilde{E}(\beta, H, 1)$, is equal to the free energy with respect to Hamiltonian H and thus $\Delta \tilde{F}$ reduces to the usual Helmholtz free energy. Conversely, for a projector onto an energy eigenstate, the effective potential, $\tilde{E}(\beta, H, |E_k\rangle\langle E_k|)$, is the corresponding energy E_k from which we regain the classical work term using a two point projective measurement scheme. More generally, when restricting to projective measurement operators, the function $\beta \tilde{E}(\beta, H, |\psi\rangle\langle\psi|)$ is a cumulant generating function in the parameter β that captures the statistical properties of measurements of H on $|\psi\rangle$ [41].

We regain the Crooks equality from this global fluctuation relation for a thermal system and a battery with a well defined energy. Specifically, in the forwards process, the system is prepared in a thermal state

$$\gamma_S^i \propto \exp\left(-\beta H_S^i\right) \tag{13}$$

and we consider the probability to observe the battery to have energy E_f having prepared it with energy E_i, that is transition probabilities of the form

$$\mathcal{P}(E_f | \gamma_S^i, E_i) := \mathcal{Q}\left(1_S \otimes |E_f\rangle\langle E_f| \,\Big|\, \gamma_S^i \otimes |E_i\rangle\langle E_i| \right). \tag{14}$$

In this classical limit, the global fluctuation relation reduces to

$$\frac{\mathcal{P}(E_f | \gamma_S^i, E_i)}{\mathcal{P}(E_i | \gamma_S^f, E_f)} = \exp\left(\beta(W - \Delta F)\right) \tag{15}$$

where $W := E_i - E_f$ is the negative change in energy of the battery and thus, due to global energy conservation, equivalent to the work done on the system. If we additionally assume that the dynamics of the system and battery do not depend on the initial energy of the battery, then using this *energy translation invariance* assumption, which we explicitly define in Section 3.3, one is able to regain all classical and semi-classical fluctuation results [29]. The global fluctuation relation is thus a genuine quantum generalisation of these relations and inherits their utility.

In this manuscript, we use the global fluctuation relation, Equation (9), to quantify deviations from the classical Crooks relation resulting from athermality of the initial thermal system and quantum coherence of the battery. Specifically, to probe the impact of preparing the system in imperfectly thermal states, we derive in Section 3.1 a Crooks-like relation for a system that is prepared in a photon added or a photon subtracted thermal state. In Section 3.2, we investigate the deviations generated by coherence in the battery by deriving a Crooks equality for binomial states of the battery.

3. Results

3.1. Photon Added and Subtracted Thermal States

Photon added and subtracted states are non-equilibrium states generated from a thermal state by, as the name suggests, either the addition or the subtraction of a single photon. Considering a single quantised field mode with creation and annihilation operators $a^\dagger$ and a and Hamiltonian H, the photon added thermal state can be written as

$$\gamma_H^+ \propto a^\dagger \exp(-\beta H) a \qquad (16)$$

and the photon subtracted thermal state as

$$\gamma_H^- \propto a \exp(-\beta H) a^\dagger . \qquad (17)$$

The states γ_H^+ and γ_H^- are diagonal in the energy eigenbasis and therefore are classical in the sense that they are devoid of coherence. Nonetheless, they are non-Gaussian and have negative Wigner functions [45,65–68], traits which are considered non-classical in the context of quantum optics.

Moreover, the addition or subtraction of a photon from a thermal state has a rather surprising impact on the number of photons in the state: In particular, adding a photon to a thermal state of light, which contains on average $\bar{n}$ photons, increases the expected number of photons in the state to $2\bar{n} + 1$ [42–44]. Similarly, subtracting a photon from a thermal state doubles the expected number of photons to $2\bar{n}$. Thus, counter-intuitively, adding or subtracting a *single* photon to a thermal state *substantially increases* the expected number of photons in the state.

In line with standard nomenclature we will refer to *photon* added and subtracted thermal states throughout this paper; however, the modes in Equations (16) and (17) could naturally refer to any boson. Experimental techniques for generating photon added [45] and subtracted [42] thermal states are well established and methods are currently being developed for the preparation of *phonon* added states [69].

To illustrate the deviations from classical thermodynamics induced by the addition (subtraction) of a single photon we derive a Crooks-like relation characterised by replacing the initially thermal system of the standard setting quantified by the Crooks equality, with a system in a photon added (subtracted) thermal state. That is, for the photon added (+) and photon subtracted (−) equalities, we suppose that the system is prepared in the states

$$\rho_S^i = \gamma_i^\pm \quad \text{and} \quad \rho_S^f = \gamma_f^\pm \qquad (18)$$

at the start of the forwards and reverse processes, respectively, where to simplify notation we have introduced the shorthand $\gamma_k^\pm \equiv \gamma_{H_S^k}^\pm$.

In analogy to the classical Crooks relation, we quantify the work supplied to the system when the photon added (subtracted) thermal system is driven by a change in Hamiltonian. For concreteness, we assume here that the system is a quantum harmonic oscillator with initial and final Hamiltonians given by

$$H_S^k := \hbar \omega_k \left(a_k^\dagger a_k + \frac{1}{2} \right), \qquad (19)$$

for $k = i$ and $k = f$, such that the system is driven by a change in its frequency from ω_i to ω_f. As energy is globally conserved, the work supplied to the system is given by the change in energy of the battery and therefore the probability distribution for the work done on the system can be quantified by transition probabilities between energy eigenstates of the battery. Specifically, in the forward process, we consider the probability to observe the battery to have energy E_f having prepared it with energy E_i and vice versa in the reverse. We do not need to make any specific assumptions on the battery Hamiltonian H_B to quantify such eigenstate transition probabilities and therefore H_B may be chosen freely.

In contrast to the usual Crooks relation, the photon added (subtracted) Crooks relations depends on the average number of photons in the photonic system after the driving process. This arises from the mapping $\mathcal{M}$ between the measurement operators and the initial states following Equation (7).

As shown explicitly in Appendix A, on inverting Equation (7), we find that for the photon added equality the measurement operators X_S^i and X_S^f are given by

$$X_S^k = a_k^\dagger a_k := N_k \quad \text{for} \quad k = i, f; \tag{20}$$

and for the subtracted equality they are given by

$$X_S^k = a_k a_k^\dagger = N_k + 1 \quad \text{for} \quad k = i, f. \tag{21}$$

That is, in both cases, they are given in terms of the number operator N_k only.

Given this form for the measurement operators, it follows that the photon added and subtracted Crooks relations quantify the expected number of photons in the system at the end of the driving process as well as the change in energy of the system. For example, for the forwards process of the photon added Crooks equality, $\mathcal{Q}$, as defined in Equation (4), is equal to

$$\mathcal{Q}\left(N \otimes |E_f\rangle\langle E_f| \, \Big| \, \gamma_i^+ \otimes |E_i\rangle\langle E_i|\right) = n(E_f|\gamma_i^+, E_i)\, \mathcal{P}(E_f|\gamma_i^+, E_i) \tag{22}$$

where $\mathcal{P}$ is the transition probability of the battery from energy E_i to E_f conditional on preparing the system in a photon added thermal state, as defined in Equation (14), and $n(E_f|\gamma_i^+, E_i)$ is the average number of photons in the system at the end of this driving process. Similar expressions to Equation (22) are obtained for the reverse process of the photon added equality and both the forwards and reverse processes of the photon subtracted equality.

As we are considering transition probabilities between energy eigenstates of the battery, the generalised energy flow term $\Delta \tilde{W}$ reduces to the work done on the system as in Equation (15). However, as derived explicitly in Appendix A, the generalisations of the free energy term, Equation (10), $\Delta \tilde{F}^+$ and $\Delta \tilde{F}^-$ for the photon added and subtracted equalities, respectively, evaluate to

$$\Delta \tilde{F}^\pm = 2\Delta F \pm \Delta E_{\text{vac}}. \tag{23}$$

In the above, ΔF is the change in free energy associated with the change in Hamiltonian from H_S^i to H_S^f and we introduce ΔE_{vac},

$$\Delta E_{\text{vac}} := \frac{1}{2}\hbar\omega_f - \frac{1}{2}\hbar\omega_i, \tag{24}$$

as the difference between the initial and final vacuum energies of photonic system.

In the classical limit where $\hbar$ tends to zero, the contribution from the energy of the vacuum state, ΔE_{vac}, vanishes and ΔF^+ and ΔF^- both tend to $2\Delta F$. This behaviour can be explained by the observation in [44] that the photon probability distributions for photon added and subtracted states have the same functional form but while the photon subtracted distribution starts at $n = 0$, that is in the vacuum state, the photon added distribution starts at $n = 1$, and therefore has no vacuum contribution, a shift which becomes increasingly insignificant for higher temperatures. Conversely, as shown in Figure 2, in the low temperature quantum limit, the contribution of the energy of the vacuum state generates sizeable deviations between the generalised free energy terms for the photon added and subtracted cases. Specifically, while ΔF_S^- tends to ΔF in agreement with the standard classical Crooks relation, we find that ΔF^+ is substantially larger than $2\Delta F$. This is due to the fact that in the low temperature limit the photon subtracted thermal state and normal thermal state both tend to the vacuum state, whereas the photon added thermal state tends to a single photon Fock state. In all limits, ΔF^+ and ΔF^- are larger than ΔF, indicating that the addition and subtraction of a photon increases the energy and information content of a thermal state, thereby increasing the extractable work from the state. Similar phenomena have been observed elsewhere in the context of work extraction protocols [46] and Maxwell demons [47].

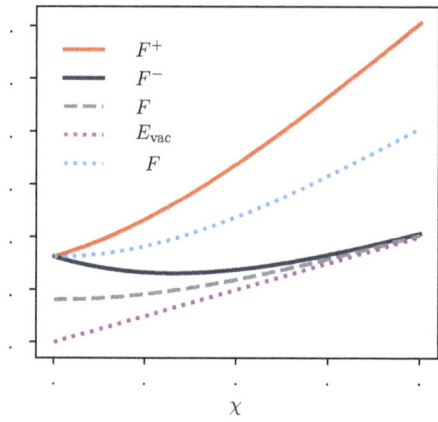

Figure 2. Generalised Free Energies. The solid red and dark blue lines show the generalised free energy, ΔF^+ and ΔF^-, of the oscillator system for the photon added and photon subtracted equalities, respectively. These are plotted as a function of $\chi = \beta\hbar\omega_i/2$, the ratio between the initial vacuum fluctuations, $\hbar\omega_i/2$, and the thermal fluctuations, $k_B T$, a measure which quantifies the temperature and thus effectively delineates the classical and quantum regimes. The grey dashed line is the usual change in energy ΔF. The dotted lines indicate the contribution of ΔE_{vac} (purple) and $2\Delta F$ (light blue) to ΔF^+ and ΔF^+. In this plot, we suppose $\hbar\omega_f = 1.5\hbar\omega_i$ and energies are given in units of $k_B T$.

The final photon added (+) and photon subtracted (−) Crooks equality can be written as

$$\frac{P(E_f|\gamma_i^\pm, E_i)}{P(E_i|\gamma_f^\pm, E_f)} = \mathcal{R}_\pm(W) \exp\left(\beta\left(W - 2\Delta F \mp \Delta E_{\text{vac}}\right)\right). \tag{25}$$

The prefactor $\mathcal{R}_\pm(W)$ quantifies the ratio of the number of photons measured in the system at the end of the reverse process over the number of photons measured at the end of the forwards process. Note, as a result, the prefactor is only defined when both the numerator and denominator of Equation (A30) are both positive quantities. As shown in Appendix A, the prefactors $\mathcal{R}_+(W)$ and $\mathcal{R}_-(W)$ can be written as

$$\mathcal{R}_\pm(W) = \frac{\omega_f}{\omega_i}\frac{\hbar\omega_f(2\bar{n}_f + k_\pm) + W + \Delta E_{\text{vac}}}{\hbar\omega_i\left(2\bar{n}_i + k_\pm^{-1}\right) - W - \Delta E_{\text{vac}}} \tag{26}$$

with $k_+ = 1$ and $k_- = \frac{\omega_i}{\omega_f}$ and where $\bar{n}_k$ is the average number of photons in a thermal state with frequency ω_k. It is worth noting that $\mathcal{R}_\pm(W)$ implicitly depends on the free energy of the initial and final Hamiltonians because $\hbar\omega_k(\bar{n}_k + \frac{1}{2})$ is the average energy of a thermal photonic state with frequency ω_k, which, by definition, is equal to the sum of free energy and entropy of the state.

The classical Crooks equality implies that driving processes which require work and decrease free energy are exponentially more likely than processes which produce work and increase free energy, thus quantifying the irreversibility of non-equilibrium driving processes. Given that the generalised free energy terms $\Delta \tilde{F}^+$ and $\Delta \tilde{F}^-$ are greater than the usual change in free energy ΔF, it is tempting to conclude that athermality of the initial system can strengthen irreversibility by amplifying the suppression factor of free energy increasing processes. However, the presence of the prefactor $\mathcal{R}$ in Equation (25), which depends on both the work done during the driving process and implicitly the initial and final free energies of the system, makes it harder to draw clear cut conclusions.

To aid comparison between the athermal and thermal cases, in Figure 3, we plot the total predicted ratio of the forwards and reverse processes for the photon added and subtracted Crooks relations, that is the right hand side of Equation (25), and compare them to the equivalent prediction of the classical relation, Equation (1). We similarly plot the prefactors $\mathcal{R}_+$ and $\mathcal{R}_-$. As the prefactor $\mathcal{R}$ does not appear in the classical Crooks relation, Equation (1), we can say that $\mathcal{R}$ is effectively equal to 1 in the limit of a perfectly thermal system. For concreteness, we here consider a forwards process where the oscillator frequency is doubled, increasing the system's free energy. We plot the ratio and $\mathcal{R}$ as a function of $\chi := \frac{\beta\hbar\omega}{2}$, the ratio of vacuum energy to thermal energy, a measure which delineates between quantum and thermodynamic regimes.

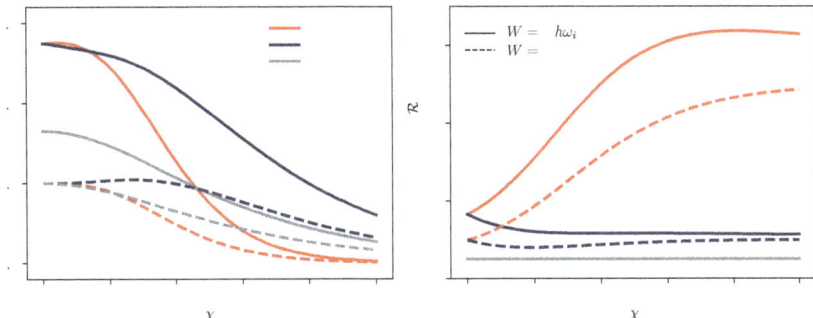

Figure 3. Predicted ratio and $\mathcal{R}$ prefactor. The left figure plots the predicted ratio of the forwards and reverse transition probabilities, i.e., the right hand side of Equation (25), for the photon added (subtracted) Crooks equality as a function of $\chi = \beta\hbar\omega_i/2$. The right figure plots $\mathcal{R}$ as a function of χ. The red (blue) lines indicates the photon added (subtracted) case and the grey lines indicate the equivalent classical limit. That is, in the left plot the grey line is the right hand side of the classical Crooks equality, Equation (1), and in the right plot the grey line is $\mathcal{R} = 1$. The solid lines plot the case $W = 2\hbar\omega_i$ and the dashed lines, $W = 0$. Here, we suppose $\hbar\omega_f = 5\hbar\omega_i$.

As shown in Figure 3, the interplay between the prefactors $\mathcal{R}_\pm$, which are greater than the classical limit of 1, and the terms $\exp(-\beta\Delta\tilde{F}_\pm)$, which are smaller than $\exp(-\beta\Delta F)$, leads to a rich spectrum of deviations from the classical Crooks relation. For example, while the prefactor $\mathcal{R}_+$ for the photon added case is substantially greater than 1 in the low temperature limit, the total predicted ratio is smaller than for the photon subtracted case. This is because the large value of $\mathcal{R}_+$ is exponentially suppressed by $\Delta\tilde{F}_+$ which is substantially larger than $\Delta\tilde{F}_-$ and ΔF, as shown in Figure 2, due to the contribution of the change in vacuum energy. Thus, we conclude that for the photon added relation, irreversibility is milder in the quantum limit due to the contribution of the energy of the vacuum state, a phenomenon which was also observed in [40].

In the high temperature classical limit one might expect adding or subtracting a single photon to a thermal state containing on average a large number of photons would have a negligible effect. Indeed, this is what we see for processes in which no work is performed on the system since in the high temperature limit the prefactor $\mathcal{R}_\pm(0)$ reduces to $\exp(\beta\Delta F)$. However, interestingly for work requiring processes, we do see large deviations from the usual classical Crooks relation in the classical limit. We attribute this to the fact that adding or subtracting a photon from thermal light effectively doubles the mean photon number the state, and therefore the net effect can be substantial even for high temperature states as they contain larger numbers of photons.

More generally, for all temperatures and for both the photon added and subtracted relations, we find that the larger the work done on the system, the larger the predicted ratio. This confirms that even when the initial states are photon added or subtracted thermal states, processes which require work are exponentially more probable than processes that generate work.

3.2. Binomial States

In the previous section, we show how the *athermality* of the initial *system*, due to the addition or subtraction of a single photon, induces rich deviations from the classical Crooks relation. Here, we complement this analysis by exploring how quantum features can be introduced through the *coherence* of the *battery*. The quantum fluctuation relations are well characterised for coherent states of the battery [40] which have close-to-classical properties. In the following, we consider binomial states, which provide a well-defined transition between coherent states of a quantum harmonic oscillator, and highly quantum mechanical states such as a state of an individual qubit.

Binomial states are pure states of the form

$$|n,p\rangle = \sum_{k=0}^{n} \sqrt{\binom{n}{k} p^k (1-p)^{n-k}} \, e^{i\phi_k} |k\rangle, \tag{27}$$

whose properties have been extensively studied in the field of quantum optics [44,53,54,70]. Binomial states are non-classical states with finite support and exhibit sub-Poissonian statistics [44,54], squeezing of quadratures [54] and are highly non-classical both in terms of their coherent properties and the negativity of their Wigner function [55]. They can be thought of as an n-qubit tensor product $|p\rangle^{\otimes n}$ of the states $|p\rangle = \sqrt{1-p}|0\rangle + \sqrt{p}|1\rangle$. The states $|n,p\rangle$ and $|p\rangle^{\otimes n}$ are related by an energy-preserving unitary rotation. This is important as the effective potential $\tilde{E}$ is invariant under energy conserving unitaries, implying that as far as the fluctuation theorem is concerned, they are interchangeable. In the limit that n tends to infinity, they approach the regular coherent states and the opposite limiting case $n=1$ corresponds to the deep quantum regime.

Binomial states find use owing to their nice analytical properties. For instance, the commonly encountered spin-coherent states are particular examples of binomial states [52,70–72]. Spin-coherent states belong to a class of generalised coherent states that allow for different displacement operators, in this case of the form $D(\alpha) = \exp(\alpha S_+ + \alpha^* S_-)$ where $S_\pm$ are the spin-raising and lowering operators [52,53,71]. Proposals for the generation of binomial states have been developed in atomic systems [56,57] and they have been suggested as analogues to coherent states for rotational systems [73,74]. These examples indicate that binomial states are of natural physical interest.

In what follows, we assume the battery is a harmonic oscillator, $H_B = \hbar\omega(a^\dagger a + \frac{1}{2})$, but do not make any specific assumptions on the initial and final system Hamiltonians. Note, one could also consider a finite Hamiltonian; however, for complete generality, decoupling the dimension of the Hamiltonian and the support of the state proves useful. We assume the system is prepared in a standard thermal state and consider transitions between two binomial states of the battery. More specifically, here the battery measurement operators are chosen as the projectors

$$X_B^k = |n_k, p_k\rangle\langle n_k, p_k| \quad \text{for} \quad k=i,f. \tag{28}$$

which, given the mapping $\mathcal{M}$ in Equation (7), fixes the preparation states. As shown in Appendix B, we find that the prepared states are the binomial states,

$$\rho_B^k = |n_k, \tilde{p}_k\rangle\langle n_k, \tilde{p}_k| \quad \text{with} \quad \tilde{p} = \frac{pe^{-\beta\hbar\omega}}{pe^{-\beta\hbar\omega} + q} \quad \text{and} \quad \tilde{q} = \frac{q}{pe^{-\beta\hbar\omega} + q}, \tag{29}$$

with $q = 1-p$ and for $k=i,f$. Thus, we see that the mapping $\mathcal{M}$ preserves binomial statistics but leads to a distortion factor due to the presence of coherence. Since $\tilde{p}$ is always less than p, this distortion from $\mathcal{M}$ lowers the energy of the prepared state as compared to the equivalent measured state, with its energy vanishing in low temperature limit.

There exist two clear distinct physical regimes corresponding to different battery preparation and measurement protocols. In the *realignment* regime, we fix the system size n and consider transition probabilities between rotated states. Conversely, the *resizing* regime quantifies transition probabilities between states of different "sizes", that is states with different supports but fixed alignment in the Bloch sphere. For the realignment regime, the prepare and measure protocols are as follows:

- Forwards: The battery B is prepared in the state $|n, \tilde{p}_i\rangle$ and measured in $|n, p_f\rangle$
- Reverse: The battery B is prepared in the state $|n, \tilde{p}_f\rangle$ and measured in $|n, p_i\rangle$.

While for the resizing regime, where we fix p and vary n, we have the prepare and measure protocol:

- Forwards: The battery B is prepared in the state $|n_i, \tilde{p}\rangle$ and measured in $|n_f, p\rangle$.
- Reverse: The battery B is prepared in the state $|n_f, \tilde{p}\rangle$ and measured in $|n_i, p\rangle$.

In the qubit picture, for a system of N qubits, the realignment regime amounts to fixing the number of battery qubits with coherence to precisely n while changing the polarisation p_k of each of these n qubits concurrently. Similarly, the resizing regime corresponds to fixing the polarisation and changing the number of non classical qubits. More precisely, we can write

$$|n_k, p_k\rangle \equiv |p_k\rangle^{\otimes n_k} \otimes |0\rangle^{\otimes N-n_k} \quad \text{for } k = i, f \tag{30}$$

where in the first regime n_k is kept fixed while p_k is varied and vice versa for the second. In the context of spin-coherent states, the first regime corresponds to a battery that remains a spin-$\frac{n}{2}$ system but whose orientation varies, while the second amounts to changing the magnitude of the spin while fixing the orientation.

The key quantity in the fluctuation relation is the generalised work flow, the derivations of which can be found in Appendix B. In these processes, the generalised work flow in the realignment regime and resizing regimes, $\Delta\tilde{W}_{\text{align}}$ and $\Delta\tilde{W}_{\text{size}}$, respectively, take the form

$$\beta\Delta\tilde{W}_{\text{align}} = n\left(\ln\frac{p_f}{\tilde{p}_f} - \ln\frac{p_i}{\tilde{p}_i}\right) \tag{31}$$

$$\beta\Delta\tilde{W}_{\text{size}} = (n_f - n_i)\left(\ln\frac{p}{\tilde{p}} + \beta\hbar\omega\right), \tag{32}$$

provided both p_i and p_f are non-zero. These capture the temperature-dependent distortion of the binomial states due to $\mathcal{M}$. While the generalised work flow in the realignment regime smoothly varies with its free parameters, in the resizing regime, the energy flow is discretised. The binomial state Crooks relations corresponding to the realignment and resizing regimes follow upon insertion of the generalised work flow terms, Equations (31) and (32), into the global fluctuation relation, Equation (9), when restricted to binomial state preparations specified in Equation (29).

In the high temperature limit, $\beta\hbar\omega \ll 1$, we can truncate the power series of $\Delta\tilde{W}$ to second order for sufficient accuracy, which gives

$$\beta\Delta\tilde{W}_{\text{align}} \approx \beta\hbar\omega n\left(p_i - p_f\right) - \frac{(\beta\hbar\omega)^2}{2}(\sigma_i^2 - \sigma_f^2) \tag{33}$$

$$\beta\Delta\tilde{W}_{\text{size}} \approx \beta\hbar\omega(n_i - n_f)p - \frac{(\beta\hbar\omega)^2}{2}(n_i - n_f)\sigma^2, \tag{34}$$

where $\sigma_k^2 = np_k(1 - p_k)$ is the variance of H_B in the state $|n, p_k\rangle$ for $k = i, f$ and $\sigma^2 = p(1 - p)$ is the variance for a Bernoulli distribution. Note that the variance evaluated for pure states is a genuine measure of coherence [75] and that due to microscopic energy conservation, that is the fact U commutes with H_{SB}, both *energy and variance in energy* are globally conserved. Given this, Equations (33) and (34) characterise the change in energy and coherence of the system due to an equal and opposite change in the battery.

Furthermore, binomial states exhibit sub-Poissonian statistics, that is the variance $np(1-p)$, is smaller than the mean np (for non vanishing p). Therefore, it follows from Equations (33) and (34) that the fluctuation relation (Equation (9)) captures the sub-Poissonian character of these states and shows that this affects the resulting irreversibility of the dynamics. Viewed through the lens of quantum optics, binomial states of light are anti-bunched [55], a signature of non-classicality. Thus, the binomial state Crooks equality draws a non-trivial link between bunching and the reversibility of quantum driving processes, since anti-bunching and sub-Poissonian statistics are directly correlated for single-mode time-independent fields [54].

In the case of spin-coherent states, the Hamiltonian is in effect taken to be defined in the eigenbasis of the spin-z operator and therefore the variances in Equations (33) and (34) detail the variation of uncertainty in the spin-z component. However, aligning the Hamiltonians in the z-direction defines a preferential axis and therefore the spin-z and the spin-x and spin-y components are not placed on equal footing. This is because the effective potential is invariant under unitary transformations U that commute with H, that is

$$E(\beta, H, \rho) = E(\beta, H, U\rho U^\dagger) \quad \forall \, [U, H] = 0 , \tag{35}$$

and hence is invariant under rotations about the z-axis. Consequently, while the fluctuation relation captures changes to the uncertainties in the spin-z components, the relation is unaffected by changes to uncertainties in the spin-x and spin-y components. More generally, the invariance of the effective potential to phase rotations means that even for standard coherent states, the fluctuation relation depends on the magnitude of the absolute displacement but not the particular magnitude of the expectation values for position and momentum. This is no coincidence, as the connection between these regimes will be explored further on.

Deviations from Classicality. To characterise the deviations between the binomial state Crooks relation and classical Crooks equality, we can compare the generalised energy flow $\Delta \tilde{W}$ to the actual energy flow in the forwards and reverse processes. In the standard Crooks equality, the work term appearing in the exponent of Equation (1) can be expressed as $W = (W - (-W))/2$, the average difference between the work done in the forward and reverse processes. For the quantum analogue, we introduce

$$W_q = (\Delta E_+ - \Delta E_-)/2 \tag{36}$$

as the difference between the energy cost ΔE_+ of the forwards process and the energy gain ΔE_- of the reverse process. Restricted to binomial state preparations of the form in Equation (29), the binomial states Crooks relation is

$$\frac{\mathcal{P}(n_f, p_f | \gamma_i; n_i, \tilde{p}_i)}{\mathcal{P}(n_i, p_i | \gamma_f; n_f, \tilde{p}_f)} = \exp\left(\beta \left(q(\chi) W_q - \Delta F\right)\right). \tag{37}$$

where the transition probabilities take the form

$$\mathcal{P}(n_f, p_f | \gamma_i; n_i, \tilde{p}_i) := \mathcal{Q}\Big(\mathbb{1} \otimes |n_f, p_f\rangle\langle n_f, p_f| \Big| \gamma_i \otimes |n_i, \tilde{p}_i\rangle\langle n_i, \tilde{p}_i| \Big) \tag{38}$$

and we introduce the quantum distortion factor

$$q(\chi) := \frac{\Delta \tilde{W}}{W_q} \tag{39}$$

as the ratio between the generalised work flow and the actual energy flows. The classical limit $q(\chi) = 1$ corresponds to a quasi-classical expression in which the quantum fluctuation relation depends only on the energy difference between the two states $|n_i, p_i\rangle$ and $|n_f, p_f\rangle$. This can be seen from Equations (33) and (34) when truncating to first order in $\beta \hbar \omega$. Deviations from unity thus capture the quantum features of the process.

The resizing and re-aligning protocols experience two related yet distinct distortions. These factors, derived in Appendix B, are

$$q_{\text{align}}(\chi) = \frac{1}{\chi} \frac{\ln(\tilde{p}_f/p_f) - \ln(\tilde{p}_i/p_i)}{(\tilde{p}_f - \tilde{p}_i) + (p_f - p_i)} \quad \text{and} \quad q_{\text{size}}(\chi) = \frac{1}{\chi} \frac{\ln(\tilde{p}/p) + 2\chi}{\tilde{p} + p} \tag{40}$$

respectively, again provided neither p_i nor p_f vanishes. These two factors are plotted in Figure 4. They are equal to each other if one of either p_i or p_f are zero, corresponding to measuring the battery in the ground state, as can be seen with the long-form equations provided in Appendix B (see Equations (A72) and (A78)).

Both factors are independent of the system size n. That only the parameter p plays a non-trivial role is relevant to the fact that it alone controls the coherent properties of binomial states. Since n is the free parameter of the resizing regime, it is particularly significant that the deviation is independent of the change in system size. Beyond this, the realignment factor is symmetric in the parameters p_i, p_f and thus does not depend on the chosen ordering of the measurements (likewise for the resizing factor with respect to n_i and n_f).

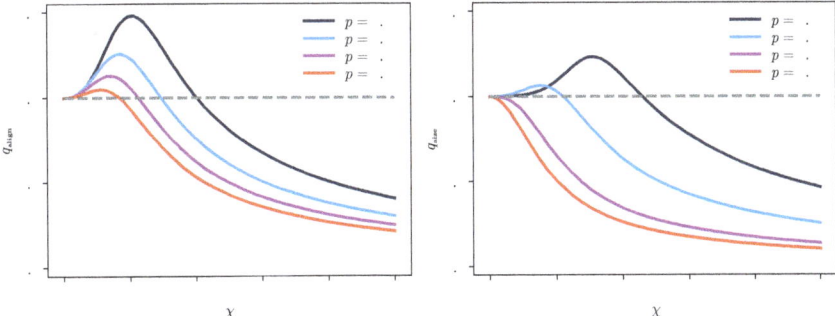

Figure 4. Quantum distortions of fluctuation relations due to binomial battery states: The left and right plots correspond to q_{align} and q_{size}, respectively. The left plot is evaluated for a fixed value $p_f = 0.8$. Both functions are plotted against the quantum-thermodynamic ratio $\chi = \frac{\beta \hbar \omega}{2}$. The plots show that the distortion due to quantum features can both enhance and suppress irreversibility in a process as compared to a "classical equivalent" solely involving energy exchanges. In both cases, we typically find suppressed irreversibility as quantum features dominate for large values of χ. However, when thermodynamic and quantum energy scales are of similar magnitude, we observe unexpected behaviour.

Regarding the thermodynamic properties, both factors exhibit a sensible classical limit in the thermally dominated regime where χ is much less than one and $\tilde{p}$ converges to p. More generally, in the quantum dominated regime at large χ, the distortion is generally sub-unity scaling as $1/\chi$, showing the irreversibility is milder than is classically expected. To understand this, consider the fact that $\tilde{E}(\beta, H, \rho)$ is lower bounded by $E^{\min}(\rho)$, defined as the smallest energy eigenvalue with non-zero weight in the state ρ [41], corresponding to the vacuum energy for all binomial states with $p < 1$. In the low temperature limit, the lower bound is saturated meaning that the generalised energy flow (accounted for by the differences in $\tilde{E}$ between any two states) vanishes. However, as shown in Figure 4, this behaviour is not true for all temperatures and values of p.

In the resizing regime, for values of p nearing unity, there exists a finite temperature region where the fluctuation relation exhibits stronger-than-classical irreversibility. Peaking for values of $p \approx 1$ in the intermediate region originate because the semi-classical two-point measurement scheme is recovered when $p = 1$, which corresponds to an energy eigenstate, hence $q_{\text{size}}(\chi) = 1$. The states satisfying this condition on p must remain close to an energy eigenstate and have a flatter initial slope until larger values of χ overcome this almost-eigenstate behaviour and recover the $1/\chi$ scaling.

The behaviour of the realigning regime is more nuanced, having two free parameters. We observe greatly enhanced irreversibility over a finite temperature region for most values of p_i or p_f if the sum of these values are $\gtrsim 1$. An oddity occurs when one measures an excited energy eigenstate, corresponding to $p_f = 1$ (due to symmetry in the parameters, one can also set $p_i = 1$ and let p_f be free). In this case, at extremely low temperatures Equation (40) is modified to

$$\lim_{\chi \to \infty} q_{\text{align}}(\chi) = \frac{2}{2-p_i} \geq 1, \qquad (41)$$

and the quantum regime no longer asymptotically approaches zero. Rather, we have that $\tilde{E}(\beta, H, \rho)$ is naturally upper bounded by $E^{\max}(\rho)$, defined as the largest eigenvalue with non-zero weight in the state ρ [41]. With the battery prepared in the excited state for either the forward or reverse protocol, we have that $\tilde{E}(\beta, H_B, |n, 1\rangle) = E^{\max}$, and the greatest possible generalised energy flow of $\Delta \tilde{W} = E^{\max} - E^{\min}$ occurs when the lower bound of $E^{\min}$ is saturated. By fixing one state to be the excited energy eigenstate, the generalised energy flow only attains this upper bound when the temperature reaches absolute zero.

At low temperatures, for values of p nearing unity the deviations from classicality are most pronounced for both regimes. Due to the temperature-dependent rescaling, this choice of parameter corresponds to the physical preparation of states with greater coherence present, as detailed by Equation (29) where p is greater than $\tilde{p}$ for all positive temperatures. Initialising the battery in a state with a large amount of coherence thus generates the non-classical behaviour we would expect.

From this analysis, we can conclude that binomial states batteries display a greater range of distinguishing features than coherent states, with the coherent properties playing a highly non-trivial role. We observe behaviour that is reminiscent of the semi-classical coherent state Crooks equality in the high and low temperature limits. In an intermediate temperature region, however, we observe deviations that lead to stronger than classical irreversibility in both the resizing and realignment regimes. We note that the binomial state factors bear many qualitative similarities to the squeezed-state factors derived in [40]. The connection between binomial and coherent states in an appropriate limit are discussed next.

The Harmonic Limit. Infinite dimensional binomial states in harmonic systems exhibit behaviour that approaches simple harmonic motion. This link is well established and leads to a semi-classical limit for the binomial state fluctuation theorem. Specifically, as shown in Appendix B, we prove that as n tends to infinity, the binomial state $|n, p\rangle$ tends to the coherent state $|\alpha\rangle$ where the displacement parameter is given by $\alpha = \sqrt{np}$ and thus is only defined as long as np remains finite. Consequently, for infinitely large spin systems, or infinitely large ensembles of qubits, with a finite expected polarisation, binomial states reduce to coherent states. Thus, in this limit, the binomial state and coherent state Crooks equalities [40] are quantitatively and qualitatively identical.

It follows that, for infinite dimensional binomial states, $q_{\text{align}}(\chi)$ and $q_{\text{size}}(\chi)$ converge on

$$q(\chi) = \frac{1}{\chi} \tanh(\chi) \qquad (42)$$

This form admits a special interpretation in terms of the mean energy of a harmonic oscillator $\hbar \omega_{\text{th}} := \langle H_B \rangle_\gamma$, with $q(\chi) = k_B T / \hbar \omega_{\text{th}}$. In particular, the average energy in a thermal harmonic oscillator is related to the thermal de Broglie wavelength λ_{th} [40]. The thermal de Broglie wavelength often finds use as a heuristic tool to differentiate between quantum and thermodynamic regimes. The coherent state equality thus leads to a natural and smooth transition between quantum and thermal properties for semi-classical battery states delineated by λ_{th}, suggesting a genuinely quantum-thermodynamic relation.

It is interesting then that the binomial state fluctuation relation is able to incorporate a wide-ranging set of features, all the way from the highly quantum single-qubit states to the semi-classical coherent state limit, together in the same framework.

3.3. Energy Translation Invariance, Jarzynski Relations and Stochastic Entropy Production

The photon added and subtracted Crooks equalities both quantify transition probabilities between states of the battery. If we assume that the system and battery dynamics depend only on the change in energy of the battery and not the initial energy of the battery, then we can rewrite the relation in terms of the probability distributions for the change in energy of the battery, that is the work done on the system. This conceptual move allows us to derive a Jarzyski-like relation for photon added and subtracted thermal states and hint at a link between the generalised free energy change and stochastic entropy production.

If the system and battery dynamics are independent of the initial energy of the battery, then the following energy translation invariance condition holds

$$\mathcal{P}(E_j|\gamma_i^\pm, E_k) = \mathcal{P}((E_j - E_k) + E_l|\gamma_i^\pm, E_l) \quad \forall\, E_j, E_k, E_l \ . \tag{43}$$

We can now define the work probability distributions in the forwards (F) and reverse (R) processes for the photon added (+) and subtracted relations (−) as

$$\mathcal{P}_F^\pm(W) := \sum_w \mathcal{P}\left(E_0 - w|\gamma_i^\pm, E_0\right) p(E_0)\, \delta(W - w) \quad \text{and} \tag{44}$$

$$\mathcal{P}_R^\pm(W) := \sum_w \mathcal{P}\left(E_0 - w|\gamma_f^\pm, E_0\right) p(E_0)\, \delta(W - w) \tag{45}$$

where $p(E_0)$ is the probability that the battery is prepared with energy E_0. It now follows, as shown in Appendix A, that the photon added and subtracted Crooks relation can be written explicitly in terms of these work distributions as

$$\frac{\mathcal{P}_F^\pm(W)}{\mathcal{P}_R^\pm(-W)} = \mathcal{R}_\pm(W) \exp\left(\beta(W \mp \Delta E_{\text{vac}} - 2\Delta F)\right) \ . \tag{46}$$

The classical Jarzynski equality, which quantifies the work done by a driven system for a *single* driving process, emerges as a corollary to the classical Crooks equality. Similarly, here, by rearranging and taking the expectation of both sides of the above equality, we obtain the photon added and subtracted Jarzynski relation

$$\left\langle \frac{1}{\mathcal{R}_\pm(W)} \exp(-\beta W) \right\rangle = \exp\left(-\beta(2\Delta F \pm \Delta E_{\text{vac}})\right) \ . \tag{47}$$

This relation complements our Crooks relation, Equation (25), by relating the work done on the athermal system for a *single* driving process, where the system's Hamiltonian is changed from H_S^i to H_S^f, to the associated change in free energy.

In classical stochastic thermodynamics, when generalising fluctuation relations to non-equilibrium initial states, such as photon added or subtracted thermal states, a natural quantity to consider is the stochastic entropy production. As expected and as shown in [31], in the limit of a classical battery which is assumed to be energy translation invariant, this inclusive setting obeys the classical Crooks equality in its formulation in terms of stochastic entropy production [9]. This suggests it may be possible to directly relate the generalised free energies term of the global fluctuation relation for non-equilibrium system states to stochastic entropy production. While these ideas were touched on in [31], explicitly stating this link remains an open question.

An analogous approach for the binomial state Crooks equality encounters difficulties. States with coherence undergo a temperature dependent rescaling and therefore the initial and final states in the forwards and reverse process are related but not equivalent. Thus, due to the presence of coherence, energy translation invariance is not a sufficient condition to rewrite the binomial state Crooks relation in terms of work probability distribution. Therefore, we cannot derive a Jarzysnki-like equality and the link with stochastic entropy production is further obscured. Similar problems arise for states such as coherent, squeezed and Schrödinger cat states, as were studied in [40].

4. Conclusions and Outlook

In this paper, we probe deviations from the classical Crooks equality induced by the initial state of the system or battery and the measurements made at the end of the driving process. However, we stress that the choice in prepared states and measurement operators is not the only manner in which the relation is non-classical. Rather, the dynamics induced by the unitary evolution will in general entangle the system and battery resulting in coherence being exchanged between the two systems. Thus, the evolved state may be a highly non-classical state. For example, for the coherent state Crooks equality, the battery is prepared in a coherent state, the most classical of the motional states of a harmonic oscillator. However, driving the battery with a change in Hamiltonian H_S^i to H_S^f, using the experimental scheme proposed in [40], results in the highly non-Gaussian state with a substantially negative Wigner function. The non-classicality of the final state can be amplified by repeating the driving process a number of times, that is cycling through changes of H_S^i to H_S^f back to H_S^i and again to H_S^f, repeatedly.

The photon added and subtracted Crooks relations could be tested by supposing that both the system and battery are photonic and using a linear optical setup, as sketched in Figure 5. Preparing a photonic battery in a high energy eigenstate, that is a Fock state containing a large but well defined number of particles, would be experimentally challenging and thus a more promising avenue is to consider a battery in a coherent state by driving one input arm with a laser. Such a scenario would be quantified by a coherent state photon added and subtracted Crooks relation. A limitation of this implementation is that it would not change the effective Hamiltonian of the system and thus only probe the relation in the limit that ΔF and ΔE_{vac} vanish. Constructing a physical implementation involving a change to the system frequency requires more imagination. One possibility would be to generalise the trapped ion implementation proposed in [40] but use a pair of internal energy levels to simulate a thermal state of an oscillator. This could be done by changing the background potential to simulate a wider range of energy level splittings.

One possible means of testing the binomial state Crooks equality would be to prepare a finite number of qubits in the state $|p\rangle = \sqrt{1-p}|0\rangle + \sqrt{p}|1\rangle$ and perform a unitary algorithm that interacts the qubits with a thermal system. This could perhaps be best performed on a quantum computer by utilising methods for Hamiltonian simulation [76,77] and with the thermal system modelled using "pre-processing" [40]. One would need to restrict to unitaries that conserve energy between the qubits and the thermal system. Both regimes could be probed with this set-up, where one could have an N qubit register and in one case prepare n_i or n_f qubits in the state $|p\rangle$, where $n_i, n_f \leq N$, or in the other case a fixed number of qubits could be individually addressed to rotate them in the Bloch sphere. Measurements in different bases are routinely performed on quantum computers and thus the measurement procedure is readily implemented.

We have taken a highly general but rather abstract fluctuation relation and shown how its physical content can be elucidated through a study of particular examples of interest. However, the cases we have considered are just a sample of the diverse range of phenomena that can be explored with this framework. While we have developed Crooks equalities for thermal systems to which a single photon has been added or subtracted, a natural extension to probe further perturbations from thermality would be to generalise our results to the case where multiple photons are added to or subtracted from the thermal state, or perhaps the case when a photon is added and then subtracted from a thermal state.

Similarly, one could quantify higher order quantum corrections to the Crooks relation by developing equalities for squeezed and cat binomial states. On a different note, incoherent binomial states, that is the dephased variant of a binomial state, model Fock states that have been transmitted through a lossy channel and thus model a lossy classical battery. Given the structural similarities between incoherent and coherent binomial states, our results here could be used to develop Crooks relations for imperfect batteries.

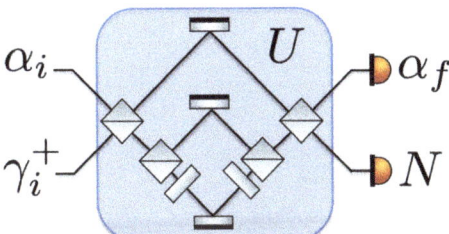

Figure 5. Linear optic implementation schematic. A photon added (or subtracted) thermal state is sent into one input arm of a linear optical set up and a coherent state the other. The linear optical set up, consisting of a series of linear optical elements, such as beamsplitters, phase-shifters and mirrors (the particular sequence sketched here is chosen arbitrarily), drives the photonic system and battery with an energy conserving and time reversal invariant operation. Finally, a coherent state measurement is performed on one output arm of the optical setup using a homodyne detection and the number of photons out put is measured in the other arm.

Author Contributions: All authors were involved with the conceptualisation of the project; the research was conducted by Z.H., E.H.M. and C.Y.-R.C.; and the manuscript was written by Z.H., E.H.M. and F.M. All authors have read and agreed to the published version of the manuscript.

Funding: This work was funded by the Engineering and Physical Sciences Research Council Centre for Doctoral Training in Controlled Quantum Dynamics and the Engineering and Physical Sciences Research Council.

Acknowledgments: We are grateful for insightful conversations with Hyukjoon Kwon.

Conflicts of Interest: The authors declare no conflict of interest. The funders had no role in the design of the study; in the collection, analyses, or interpretation of data; in the writing of the manuscript, or in the decision to publish the results.

Appendix A. Derivation of Photon Added and Subtracted Crooks Equality

The photon added (subtracted) Crooks equality is derived from the global fluctuation relation by supposing that the system is prepared in a photon added (subtracted) thermal state. That is, for the photon added (+) and photon subtracted (−) equalities, we suppose that the system is prepared in the states

$$\rho_S^i = \gamma_{H_S^i}^{\pm} \quad \text{and} \quad \rho_S^f = \gamma_{H_S^f}^{\pm} \tag{A1}$$

for the forwards and reverse processes, respectively, where the photon added state and subtracted states are defined as

$$\gamma_H^+ \propto a^\dagger \exp(-\beta H) a \quad \text{and} \quad \gamma_H^- \propto a \exp(-\beta H) a^\dagger \tag{A2}$$

respectively. In what follows, we use the short hand $\gamma_i^\pm \equiv \gamma_{H_S^i}^\pm$ and $\gamma_f^\pm \equiv \gamma_{H_S^f}^\pm$ to simplify notation. For concreteness, we consider a quantum harmonic oscillator system with initial and final Hamiltonians given by

$$H_S^k := \hbar\omega_k \left(a_k^\dagger a_k + \frac{1}{2}\right), \tag{A3}$$

for $k = i$ and $k = f$, such that the system is driven by a change in its frequency from ω_i to ω_f.

We leave the battery Hamiltonian H_B entirely general and, to isolate the deviations to the classical Crooks equality due to the athermality of the initial system states, we consider a semi-classical battery, which is prepared and measured in the energy eigenbasis. Specifically, we assume that

$$\rho_B^i = |E_i\rangle\langle E_i| \quad \text{and} \quad \rho_B^f = |E_f\rangle\langle E_f| \tag{A4}$$

where $|E_i\rangle$ and $|E_f\rangle$ are energy eigenstates of H_B. Given that the battery is prepared in energy eigenstates, the measurement operators X_B^i and X_B^f specified by Equation (7) are also projectors onto energy eigenstates, that is

$$X_B^i = |E_i\rangle\langle E_i| \quad \text{and} \quad X_B^f = |E_f\rangle\langle E_f|. \tag{A5}$$

It follows that the generalised energy flow $\Delta\tilde{E}$, Equation (11), evaluates the change in energy of the battery,

$$\Delta\tilde{W} = E_i - E_f \equiv W, \tag{A6}$$

which by global energy conservation is equivalent to the work done, W, on the system.

To derive the photon added and subtracted Crooks relations from the global fluctuation, we need to determine the measurement operators X_S^i and X_S^f which are related to the initial photon added and subtracted states by the mapping $\mathcal{M}$, Equation (7). Specifically, inverting Equation (7), we have that the measurement operators for the photon added, X_S^{k+}, and subtracted, X_S^{k-}, cases, respectively, are related to the photon added and subtracted thermal states by

$$X_S^{k\pm} \propto \exp\left(\chi_k a_k^\dagger a_k\right) \gamma_k^\pm \exp\left(\chi_k a_k^\dagger a_k\right) \tag{A7}$$

where $\chi_k = \frac{\beta\hbar\omega_k}{2}$. On substituting in the explicit expressions for γ_k^+ and γ_k^-, and using the Hadamard lemma, we find that

$$X_S^{k+} \propto \exp\left(\chi_k a_k^\dagger a_k\right) a_k^\dagger \exp\left(-2\chi_k a_k^\dagger a_k\right) a_k \exp\left(\chi_k a_k^\dagger a_k\right) \propto a_k^\dagger a_k \quad \text{and similarly,} \tag{A8}$$

$$X_S^{k-} \propto \exp\left(\chi_k a_k^\dagger a_k\right) a_k \exp\left(-2\chi_k a_k^\dagger a_k\right) a_k^\dagger \exp\left(\chi_k a_k^\dagger a_k\right) \propto a_k a_k^\dagger. \tag{A9}$$

We note that any constants of proportionality in front of the measurement operators X_S^i and X_S^f will cancel out in the final relation and thus we are free to set them to 1. We therefore conclude that the measurement operators for the photon added Crooks relation, forced by the mapping $\mathcal{M}$, Equation (7), are given by

$$X_S^{i+} = a_i^\dagger a_i \equiv N_i \quad \text{and} \quad X_S^{f+} = N_f \tag{A10}$$

and the measurement operators for the photon subtracted equality are equal to

$$X_S^{i-} = a_i a_i^\dagger = N_i + 1 \quad \text{and} \quad X_S^{f-} = N_f + 1 \tag{A11}$$

where N_i and N_f are the initial and final number operators, respectively.

The photon added Crooks equality thus quantifies the ratio of

$$\mathcal{Q}\left(a_f^\dagger a_f \otimes |E_f\rangle\langle E_f| \Big| \gamma_i^+ \otimes |E_i\rangle\langle E_i|\right) = n(E_f|\gamma_i^+, E_i)\mathcal{P}(E_f|\gamma_i^+, E_i) \tag{A12}$$

for a forwards process, and

$$\mathcal{Q}\left(a_i^\dagger a_i \otimes |E_i\rangle\langle E_i| \Big| \gamma_f^+ \otimes |E_f\rangle\langle E_f|\right) = n(E_i|\gamma_f^+, E_f)\mathcal{P}(E_i|\gamma_f^+, E_f) \tag{A13}$$

of a reverse process. Here, $n(E_f|\gamma_i^+, E_i)$ $(n(E_i|\gamma_f^+, E_f))$ is the average number of photons measured in the system at the end of the forwards (reverse) process, conditional on the battery being measured to have the energy E_f (E_i). Similarly, the photon subtracted Crooks equality quantifies the ratio of

$$\mathcal{Q}\left((a_k^\dagger a_k + 1) \otimes |E_k\rangle\langle E_k| \Big| \gamma_j^+ \otimes |E_j\rangle\langle E_j|\right) = \left(n(E_k|\gamma_j^+, E_j) + 1\right) \mathcal{P}(E_k|\gamma_j^+, E_j) \tag{A14}$$

for a forwards process, with $j = i$ and $k = f$, and a reverse process, with $j = f$ and $k = i$.

It remains to calculate the generalised free energy $\Delta \tilde{F}$ for the measurements X_S^i and X_S^f as defined in Equation (A11). To do so, we start by noting that $\Delta \tilde{F}$ can be written as

$$\Delta \tilde{F} = k_B T \ln \left(\frac{\tilde{Z}\left(\beta, H_S^i, X_S^i\right)}{\tilde{Z}\left(\beta, H_S^f, X_S^f\right)} \right) \quad \text{where} \quad \tilde{Z}(\beta, H, X) := \text{Tr}[\exp(-\beta H) X]. \tag{A15}$$

As our notation suggests, $\tilde{Z}$ is an operator dependent mathematical generalisation of the usual thermodynamic partition function,

$$Z(\beta, H_S^k) := \text{Tr}[\exp(-\beta H_S^k)]. \tag{A16}$$

For the oscillator Hamiltonians defined in Equation (A3), we find by working in the number basis that

$$\tilde{Z}\left(\beta, H_S^k, N_k\right) = \sum_{n_k=0}^\infty n_k \exp(-2\chi_k(n_k + 1/2)) = \frac{\exp(\chi_k)}{(\exp(2\chi_k) - 1)^2} \quad \text{and}$$

$$\tilde{Z}\left(\beta, H_S^k, N_k + 1\right) = \sum_{n_k=0}^\infty (n_k + 1) \exp(-2\chi_k(n_k + 1/2)) = \frac{\exp(3\chi_k)}{(\exp(2\chi_k) - 1)^2}. \tag{A17}$$

The physical content of these expressions can be elucidated by rewriting them in terms of the usual partition function, which evaluates to

$$Z(\beta, H_S^k) = \frac{\exp(\chi_k)}{\exp(2\chi_k) - 1}. \tag{A18}$$

On substituting Equation (A18) into Equation (A17), we obtain

$$\tilde{Z}\left(\beta, H_S^k, N_k\right) := Z_k \frac{1}{\exp(2\chi_k) - 1} = (Z_k)^2 \exp(-\chi_k) \quad \text{and}$$

$$\tilde{Z}\left(\beta, H_S^k, N_k + 1\right) = Z_k \frac{\exp(2\chi_k)}{\exp(2\chi_k) - 1} = (Z_k)^2 \exp(\chi_k) \tag{A19}$$

where we introduce the short hand $Z_k \equiv Z(\beta, H_S^k)$. Finally, on substituting Equation (A19) into Equation (A15), and using the fact that because

$$\frac{Z_f}{Z_i} = \exp(-\Delta F/k_B T) \quad \text{it follows that} \quad \left(\frac{Z_f}{Z_i}\right)^2 = \exp(-2\Delta F/k_B T) \tag{A20}$$

we find that

$$\Delta \tilde{F}^\pm = \pm \Delta E_{\text{vac}} + 2\Delta F. \tag{A21}$$

In the above, we introduce ΔE_{vac} as the difference between the vacuum energies of the harmonic oscillator at the start and end of the forwards driving process,

$$\Delta E_{\text{vac}} := \frac{1}{2}\hbar\omega_f - \frac{1}{2}\hbar\omega_i \ . \tag{A22}$$

The photon added (+) and photon subtracted (−) Crooks equality can thus be written as

$$\frac{P(E_f|\gamma_i^\pm, E_i)}{P(E_i|\gamma_f^\pm, E_f)} = \mathcal{R}_\pm \exp\left(\beta(W \mp \Delta E_{\text{vac}} - 2\Delta F)\right) , \tag{A23}$$

where the prefactors $\mathcal{R}_+$ and $\mathcal{R}_-$ are defined as

$$\mathcal{R}_+ := \frac{n(E_i|\gamma_f^+, E_f)}{n(E_f|\gamma_i^+, E_i)} \quad \text{and} \quad \mathcal{R}_- := \frac{n(E_i|\gamma_f^-, E_f) + 1}{n(E_f|\gamma_i^-, E_i) + 1} \ . \tag{A24}$$

Since the number of photons in the system is necessarily a positive quantity, the prefactors are only defined when both the numerator and denominator of Equation (A24) are positive quantities.

The physical role of the $\mathcal{R}_\pm$ term can be made more explicit by taking advantage of that fact that energy is conserved during the driving process. It follows that the number of photons at the end of the driving process is equal to the average number of photons initially in the system plus (minus) the change in photon number due to the decrease (increase) in the energy of the battery. By energy conservation, we can write

$$\hbar\omega_f\left(n(E_f|\gamma_i^\pm, E_i) + \frac{1}{2}\right) = \hbar\omega_i\left(n_i^\pm + \frac{1}{2}\right) - W \quad \text{and} \quad \hbar\omega_i\left(n(E_i|\gamma_f^\pm, E_f) + \frac{1}{2}\right) = \hbar\omega_f\left(n_f^\pm + \frac{1}{2}\right) + W \tag{A25}$$

where $n_i^\pm$ ($n_f^\pm$) is the average number of photons in a photon added/subtracted thermal state with frequency ω_i (ω_f) at temperature T. Equation (A25) can be rearranged to find the average number of photons measured at the end of the driving processes,

$$\hbar\omega_f n(E_f|\gamma_i^\pm, E_i) = \hbar\omega_i n_i^\pm - W - \Delta E_{\text{vac}} \quad \text{and} \quad \hbar\omega_i n(E_i|\gamma_f^\pm, E_f) = \hbar\omega_f n_f^\pm + W + \Delta E_{\text{vac}}. \tag{A26}$$

Thus, on substituting Equation (A26) into Equation (A24), we find that the prefactor $\mathcal{R}_\pm$ takes the form

$$\mathcal{R}_\pm(W) = \frac{\omega_f}{\omega_i} \frac{\hbar\omega_f n_f^\pm + W + x_\pm}{\hbar\omega_i n_i^\pm - W \mp x_\pm} \tag{A27}$$

with x_+ equal to the *change* in vacuum energy, $x_+ = \Delta E_{\text{vac}}$, and x_- equal to the sum of the initial and final vacuum energies, $x_- = \frac{\hbar\omega_f + \hbar\omega_i}{2}$. As discussed in Section 3.1, the average number of photons in a photon added or subtracted state, $n_f^\pm$, evaluates to

$$n_k^+ = 2\bar{n}_k + 1 \quad \text{and} \quad n_k^- = 2\bar{n}_k \tag{A28}$$

where $\bar{n}_k$ is the average number of photons in a thermal state with frequency ω_k and takes the form

$$\bar{n}_k := \frac{1}{Z_k}\sum n_k \exp(-2\chi_k(n_k + 1/2)) = \frac{1}{\exp(2\chi_k) - 1} \ . \tag{A29}$$

Thus, we find that the prefactor $\mathcal{R}_\pm$, Equation (A27), can be rewritten in terms of the mean number of photons in a thermal state as

$$\mathcal{R}_\pm(W) = \frac{\omega_f}{\omega_i} \frac{\hbar\omega_f(2\bar{n}_f + k_\pm) + W + \Delta E_{\text{vac}}}{\hbar\omega_i\left(2\bar{n}_i + k_\pm^{-1}\right) - W - \Delta E_{\text{vac}}} \tag{A30}$$

with $k_+ = 1$ and $k_- = \frac{\omega_i}{\omega_f}$. It is worth noting that the prefactor implicitly depends on the free energy of the initial and final Hamiltonians because the term $\hbar\omega_k(\bar{n}_k + \frac{1}{2})$ is the average energy of a photon in a thermal state with frequency ω_k, which is equal to the free energy of the state plus k_BT times the entropy of the state. Thus, $\mathcal{R}$ depends on the temperature, the work done during the driving process, the equilibrium free energy and the entropy of a thermal system with respect to the initial and final Hamiltonians.

Photon added and subtracted Jarzynski equality.

We can derive a Jarzyski-like relation for photon added and subtracted thermal states from Equation (25), if we further assume that the system and battery dynamics depend only on the change in energy of the battery and not the initial energy of the battery. That is, if the following energy translation invariance condition holds,

$$\mathcal{P}(E_j|\gamma_i^\pm, E_k) = \mathcal{P}((E_j - E_k) + E_l|\gamma_i^\pm, E_l) \quad \forall \; E_j, E_k, E_l \; . \tag{A31}$$

Having made this assumption, we can rewrite the photon added and subtracted Crooks relation, Equation (A23), as

$$\frac{\mathcal{P}(w + E_0|\gamma_i^\pm, E_0)}{\mathcal{P}(-w + E_0|\gamma_f^\pm, E_0)} = \mathcal{R}_\pm(w) \exp\left(\beta(w \mp \Delta E_{\text{vac}} - 2\Delta F)\right) , \tag{A32}$$

which can be rearranged into

$$\frac{1}{\mathcal{R}_\pm(w)} \exp(-\beta w) \mathcal{P}(w + E_0|\gamma_i^\pm, E_0) p(E_0) = \exp\left(\beta(\mp\Delta E_{\text{vac}} - 2\Delta F)\right) \mathcal{P}(-w + E_0|\gamma_f^\pm, E_0) p(E_0) \tag{A33}$$

where $p(E_0)$ is the probability that the battery is prepared with energy E_0. We can now define the work probability distributions in the forwards (F) and reverse (R) processes for the photon added (+) and subtracted relations (−) as

$$\mathcal{P}_F^\pm(W) := \sum_w \mathcal{P}\left(E_0 - w|\gamma_i^\pm, E_0\right) p(E_0) \delta(W - w) \quad \text{and} \tag{A34}$$

$$\mathcal{P}_R^\pm(W) := \sum_w \mathcal{P}\left(E_0 - w|\gamma_f^\pm, E_0\right) p(E_0) \delta(W - w) \; . \tag{A35}$$

It therefore follows from Equation (A33) that the photon added and subtracted Crooks equalities can be rewritten in terms of the forwards and reverse work probability distributions instead of battery state transition probabilities, with

$$\frac{\mathcal{P}_F^\pm(W)}{\mathcal{P}_R^\pm(-W)} = \mathcal{R}_\pm(W) \exp\left(\beta(W \mp \Delta E_{\text{vac}} - 2\Delta F)\right) \; . \tag{A36}$$

Finally, rearranging and taking the expectation of both sides of the above equality we obtain the photon added and subtracted Jarzynski relation

$$\left\langle \frac{1}{\mathcal{R}_\pm(W)} \exp(-\beta W) \right\rangle = \exp\left(-\beta(2\Delta F \pm \Delta E_{\text{vac}})\right) \; . \tag{A37}$$

Thus, we can relate the work done on a system which is prepared in a photon added or subtracted thermal state and driven by a change in Hamiltonian to the change in free energy associated with the change in Hamiltonian.

Appendix B. Derivation of Binomial State Properties

This section contains derivations of some mathematical properties of binomial states. In the main text, the mapping $\mathcal{M}$ is introduced, defined as

$$\mathcal{M}(X) = \frac{\mathcal{T}\left(e^{-\frac{\beta H_B}{2}} X e^{-\frac{\beta H_B}{2}}\right)}{\text{Tr}(e^{-\beta H} X)}. \tag{A38}$$

This mapping, without the time-reversal, is often referred to as a Gibbs rescaling, and it has many interesting properties [29,41]. Under the Gibbs rescaling, we find the binomial states transform as follows.

Proposition 1. *Let $|n, p\rangle$ be a binomial state as defined in the main text, for $n \in \mathbb{N}$ and $0 \leq p \leq 1$. For a harmonic Hamiltonian $H_B = \hbar\omega(a^\dagger a + \frac{1}{2})$, the Gibbs re-scaled state $|n, \tilde{p}\rangle\langle n, \tilde{p}| = \Gamma_{H_B}(|n, p\rangle\langle n, p|)$ is also a binomial state with probability distribution*

$$\tilde{p} := \frac{e^{-\beta\hbar\omega} p}{pe^{-\beta\hbar\omega} + q}, \quad \tilde{q} := \frac{q}{pe^{-\beta\hbar\omega} + q}, \tag{A39}$$

where $q = 1 - p$.

Proof. Since the Gibbs re-scaling maps pure states to pure states, we need only consider the action of $\mathcal{Z}_{n,p}^{-1/2} e^{-\beta H_B/2} |n, p\rangle = |\psi\rangle$ where $\mathcal{Z}_{n,p}^{-1/2}$ is the normalising factor. As the phases are arbitrary, we neglect them with no loss of generality. Before proceeding, we make the substitution $\chi = \frac{\beta\hbar\omega}{2}$ and $q = 1 - p$. Using the definition of $|n, p\rangle$, we find

$$|\psi\rangle = \frac{1}{\sqrt{\mathcal{Z}_{n,p}}} \sum_{k=0}^{n} \sqrt{\binom{n}{k} p^k q^{n-k}} e^{-\chi(a^\dagger a + \frac{1}{2})} |k\rangle \tag{A40}$$

$$= \frac{1}{\sqrt{\mathcal{Z}_{n,p}}} \sum_{k=0}^{n} \sqrt{\binom{n}{k} p^k q^{n-k}} e^{-\chi(k+\frac{1}{2})} |k\rangle. \tag{A41}$$

Let us calculate the normalisation factor

$$\mathcal{Z}_{n,p} = \langle n, p | e^{-\beta H_S} | n, p\rangle \tag{A42}$$

$$= \sum_{k=0}^{n} \frac{n!}{k!(n-k)!} p^k q^{n-k} e^{-2\chi(k+\frac{1}{2})} \tag{A43}$$

$$= e^{-\chi}(pe^{-2\chi} + q)^n \tag{A44}$$

where to obtain the last line we used the binomial expansion theorem. Inserting Equation (A44) into Equation (A41), we obtain

$$|\psi\rangle = \frac{e^{-\chi/2}}{e^{-\chi/2}(pe^{-2\chi} + q)^{n/2}} \sum_{k=0}^{n} \sqrt{\binom{n}{k} p^k q^{n-k}} e^{-k\chi} |k\rangle \tag{A45}$$

$$= \sum_{k=0}^{n} \sqrt{\binom{n}{k} \frac{p^k q^{n-k}}{(pe^{-2\chi} + q)^n}} e^{-2k\chi} |k\rangle \tag{A46}$$

$$= \sum_{k=0}^{n} \sqrt{\binom{n}{k} \left[\frac{pe^{-2\chi}}{pe^{-2\chi} + q}\right]^k \left[\frac{q}{pe^{-2\chi} + q}\right]^{n-k}} |k\rangle \tag{A47}$$

$$= \sum_{k=0}^{n} \sqrt{\binom{n}{k} \tilde{p}^k \tilde{q}^{n-k}} |k\rangle, \tag{A48}$$

where
$$\tilde{p} := \frac{e^{-\beta\hbar\omega}p}{pe^{-\beta\hbar\omega} + q}, \quad \tilde{q} := \frac{q}{pe^{-\beta\hbar\omega} + q}. \tag{A49}$$

It is easily verified that $\tilde{p} + \tilde{q} = 1$ and therefore $|\psi\rangle = |n, \tilde{p}\rangle$ is a binomial state as claimed. □

Binomial state statistics are preserved under a Gibbs re-scaling but in general $\tilde{p}$ decreases with increasing χ, as can be seen if we instead look at $\tilde{q}$. In the limit $\chi \to 0$, $\tilde{q} \to q$ and hence $\tilde{p} \to p$, while, in the limit $\chi \to \infty$, $\tilde{q} \to 1$ and conversely $\tilde{p} \to 0$. It smoothly varies between these two limits, implying $\tilde{q} \geq q$.

To derive the quantum distortion factors, we need to know the expectation value in energy for a system prepared in a binomial state.

Proposition 2. *Suppose B has a harmonic Hamiltonian $H_B := \hbar\omega(a^\dagger a + \frac{1}{2})$; then, the expectation value of energy for a state $|n, p\rangle$ is*
$$\langle H_B \rangle_{n,p} = \hbar\omega \left(np + \frac{1}{2} \right). \tag{A50}$$

Proof. We begin by assuming a harmonic Hamiltonian $H_B = \hbar\omega(a^\dagger a + \frac{1}{2})$. Using the definition of $|n, p\rangle$ leads to
$$\langle H_B \rangle_{n,p} = \sum_{k=0}^{n} \frac{n!}{k!(n-k)!} p^k q^{n-k} \hbar\omega \left(k + \frac{1}{2} \right) \tag{A51}$$

We now proceed to calculate the two components separately; for the first, we have
$$\text{first term} = \hbar\omega n p \sum_{k=0}^{n} k \frac{(n-1)!}{k!(n-k)!} p^{k-1} q^{n-k} \tag{A52}$$
$$= \hbar\omega n p \sum_{k=1}^{n} \frac{(n-1)!}{(k-1)!([n-1]-[k-1])!} p^{k-1} q^{[n-1]-[k-1]} \tag{A53}$$
$$= \hbar\omega n p \sum_{j=0}^{m} \frac{m!}{j!(m-j)!} p^j q^{m-j} \tag{A54}$$
$$= \hbar\omega n p (p+q)^m \tag{A55}$$
$$= \hbar\omega n p \tag{A56}$$

where we made the substitutions $m = n - 1$ and $j = k - 1$. Doing a similar calculation for the second term,
$$\text{second term} = \frac{\hbar\omega}{2} \sum_{k=0}^{n} \frac{n!}{k!(n-k)!} p^k q^{n-k} \tag{A57}$$
$$= \frac{\hbar\omega}{2} (p+q)^n \tag{A58}$$
$$= \frac{\hbar\omega}{2}. \tag{A59}$$

Combining these two equations gives the claimed result. □

The final property we need is the effective potential evaluated for an arbitrary binomial state

Proposition 3. *For a binomial state $|n, p\rangle$ and harmonic Hamiltonian $H_B = \hbar\omega(a^\dagger a + \frac{1}{2})$,*
$$\beta \tilde{E}(\beta, H_B, |n, p\rangle) = \frac{\beta\hbar\omega}{2} - n \ln(pe^{-\beta\hbar\omega} + q) \tag{A60}$$

where $q = 1 - p$.

Proof. From the definition of $\tilde{E}$ and $|n, p\rangle$, we find

$$\beta \tilde{E}(\beta, H_B, |n, p\rangle) = -\ln \left(\sum_{k=0}^{n} e^{-\beta \hbar \omega (k + \frac{1}{2})} \binom{n}{k} p^k q^{n-k} \right) \tag{A61}$$

$$= -\ln \left(e^{-\beta \hbar \omega / 2} \sum_{k=0}^{n} e^{-\beta \hbar \omega k} \binom{n}{k} p^k q^{n-k} \right) \tag{A62}$$

$$= \frac{\beta \hbar \omega}{2} - \ln \left(\sum_{k=0}^{n} \binom{n}{k} \left[p e^{-\beta \hbar \omega} \right]^k q^{n-k} \right) \tag{A63}$$

$$= \frac{\beta \hbar \omega}{2} - \ln \left(p e^{-\beta \hbar \omega} + q \right)^n \tag{A64}$$

$$= \frac{\beta \hbar \omega}{2} - n \ln \left(p e^{-\beta \hbar \omega} + q \right), \tag{A65}$$

which concludes the proof. □

Equation (A60) can also be formulated in terms of $\tilde{p}$ by noting that $p e^{-\beta \hbar \omega} / \tilde{p} = (p e^{-\beta \hbar \omega} + q)$. It follows that

$$\beta \tilde{E}_B(\beta, H_B, |n, p\rangle) = \frac{\beta \hbar \omega}{2} - n \ln(p e^{-\beta \hbar \omega} / \tilde{p}) \tag{A66}$$

$$= \beta \hbar \omega (n + \frac{1}{2}) + n \ln(\tilde{p}/p), \tag{A67}$$

on the condition that $p, \tilde{p} > 0$.

Appendix B.1. The Quantum Distortion Factor for Binomial States

In the main text, we discuss a quantum distortion factor $q(\chi)$ that determines how the quantum fluctuation theorem diverges compared to the standard notion of average change in energy of the forwards and reverse processes. Here, we derive the explicit formulae for $q(\chi)$.

We defined two distinct processes when restricting to binomial state preparation and measurement, corresponding to the *resizing* and *re-aligning* regimes. In the re-aligning regime, using Proposition 2, the energetic *cost to the battery* in each protocol is

$$\Delta E_+^{(\text{align})} := \langle H_B \rangle_{n,\tilde{p}_i} - \langle H_B \rangle_{n,p_f} = \hbar \omega n (\tilde{p}_i - p_f), \tag{A68}$$

$$\Delta E_-^{(\text{align})} := \langle H_B \rangle_{n,\tilde{p}_f} - \langle H_B \rangle_{n,p_i} = \hbar \omega n (\tilde{p}_f - p_i). \tag{A69}$$

The quantity $W_q^{(\text{align})} := (\Delta E_+^{(\text{align})} - \Delta E_-^{(\text{align})})/2$ therefore takes the form

$$W_q^{(\text{align})} = \frac{\hbar \omega n}{2} \left([\tilde{p}_i + p_i] - [\tilde{p}_f + p_f] \right) = \frac{\hbar \omega}{2} \left(p_i \left[\frac{e^{-\beta \hbar \omega}}{p_i e^{-\beta \hbar \omega} + q_i} + 1 \right] - p_f \left[\frac{e^{-\beta \hbar \omega}}{p_f e^{-\beta \hbar \omega} + q_f} + 1 \right] \right). \tag{A70}$$

For the generalised energy flow, we can use Proposition 3, which depends solely upon the normal un-rescaled states $\Delta \tilde{W} = E(\beta, H_B, |n, p_i\rangle) - E(\beta, H_B, |n, p_f\rangle)$. This turns out to be

$$\Delta \tilde{W}_{\text{align}} = -n k_B T \ln \left(\frac{p_i e^{-\beta \hbar \omega} + q_i}{p_f e^{-\beta \hbar \omega} + q_f} \right) = -n k_B T \ln \left(\frac{\tilde{p}_f p_i}{p_f \tilde{p}_i} \right), \tag{A71}$$

where the latter equality holds provided $p_f, p_i \neq 0$. The quantum distortion factor where we are free to vary p for fixed n thus takes the form

$$q_{\text{align}}(\chi) = \frac{1}{\chi} \frac{\ln\left(\frac{p_i e^{-2\chi+q_i}}{p_f e^{-2\chi+q_f}}\right)}{(\tilde{p}_f - \tilde{p}_i) + (p_f - p_i)} \tag{A72}$$

$$= \frac{1}{\chi} \frac{\ln\left(\frac{\tilde{p}_f}{p_f}\right) - \ln\left(\frac{\tilde{p}_i}{p_i}\right)}{(\tilde{p}_f - \tilde{p}_i) + (p_f - p_i)}, \quad \text{if } p_i, p_f \neq 0. \tag{A73}$$

On the other hand, one is also free to vary n and keep p fixed as detailed by the resizing regime. We can define the same quantities, which we now label with a new superscript to differentiate the cases.

$$\Delta E_+^{(\text{size})} := \langle H_B \rangle_{n_i, \tilde{p}} - \langle H_B \rangle_{n_f, p} = \hbar\omega(n_i \tilde{p} - n_f p), \tag{A74}$$

$$\Delta E_-^{(\text{size})} := \langle H_B \rangle_{n_f, \tilde{p}} - \langle H_B \rangle_{n_i, p} = \hbar\omega(n_f \tilde{p} - n_i p), \tag{A75}$$

which implies

$$W_q^{(\text{size})} = \frac{\hbar\omega}{2}\left(n_i - n_f\right)(\tilde{p} + p) = \frac{\hbar\omega p}{2}\left(n_i - n_f\right)\left(\frac{e^{-\beta\hbar\omega}}{pe^{-\beta\hbar\omega} + q} + 1\right). \tag{A76}$$

Likewise, the generalised energy flow for this process is given by

$$\Delta \tilde{W}_{\text{size}} = -(n_i - n_f)k_B T \ln(pe^{-\beta\hbar\omega} + q) = (n_i - n_f)\{k_B T \ln(\tilde{p}/p) + \beta\hbar\omega\}. \tag{A77}$$

The quantum distortion factor for the second regime is thus

$$q_{\text{size}}(\chi) = \frac{1}{\chi} \frac{\ln(pe^{-2\chi} + q)}{\tilde{p} + p} \tag{A78}$$

$$= \frac{1}{\chi} \frac{\ln(\tilde{p}/p) + 2\chi}{\tilde{p} + p}, \quad \text{if } p \neq 0. \tag{A79}$$

Appendix B.2. The Harmonic Limit

In this section, we prove that there exists a limit in which binomial states become coherent states with arbitrary precision. In what follows, we assume that $np = \lambda$ for some constant $\lambda \in \mathbb{R}$. The correct limit involves making the binomial states a superposition over infinitely many energy eigenstates by taking $n \to \infty$ and correspondingly $p \to 0$.

Firstly, let us consider the effect on the expectation value for energy. We have that

$$\lim_{\substack{n \to \infty \\ np = \lambda}} \langle H_B \rangle_{n,p} = \lim_{\substack{n \to \infty \\ np = \lambda}} \hbar\omega\left(np + \frac{1}{2}\right) \tag{A80}$$

$$= \hbar\omega\left(\lambda + \frac{1}{2}\right) \tag{A81}$$

which we note bears a likeness to the expectation value of energy for a coherent state $|\alpha\rangle$ where $|\alpha|^2 = \lambda$. Likewise, the effective potential also attains an identical form to that of a coherent state $\tilde{E}(\beta, H_B, |\alpha\rangle)$ where we once again choose $|\alpha|^2 = \lambda$.

$$\lim_{\substack{n\to\infty\\ np=\lambda}} \beta \tilde{E}_B(\beta, |n, p\rangle) = \lim_{\substack{n\to\infty\\ np=\lambda}} \left(\frac{\beta\hbar\omega}{2} - n\ln\left(1 + \frac{\lambda}{n}[e^{-\beta\hbar\omega} - 1]\right)\right) \tag{A82}$$

$$= \lim_{\substack{n\to\infty\\ np=\lambda}} \left(\frac{\beta\hbar\omega}{2} - n\left[\frac{\lambda}{n}[e^{-\beta\hbar\omega} - 1] + \mathcal{O}\left(\frac{1}{n^2}\right)\right]\right) \tag{A83}$$

$$= \frac{\beta\hbar\omega}{2} + \lambda(1 - e^{-\beta\hbar\omega}). \tag{A84}$$

For our purposes, these two quantities being identical to their coherent state counterparts means that the fluctuation theorem in the appropriate limit is indistinguishable from a coherent state fluctuation theorem. However, it is also the case that the states themselves become identical. This is easily verified by using the closely related characteristic functions [78]. Since characteristic functions $\varphi(t)$ uniquely specify a probability distribution, showing equality for all t translates to equality in distribution. Defining the characteristic function $\varphi_\psi(t) := \langle \psi | e^{iH_B t} | \psi \rangle$ we have

$$\varphi_\alpha(t) = e^{|\alpha|^2(e^{i\hbar\omega t} - 1) + i\frac{\hbar\omega}{2} t} \tag{A85}$$

$$\varphi_{n,p}(t) = e^{i\frac{\hbar\omega}{2} t}(1 + p[e^{i\hbar\omega t} - 1])^n. \tag{A86}$$

Making the substitution $p = \lambda/n$, we find

$$\varphi_{n,p}(t) = e^{i\frac{\hbar\omega}{2} t}\left(1 + \frac{\lambda}{n}[e^{i\hbar\omega t} - 1]\right)^n \tag{A87}$$

However, in the limit, we have that $\lim_{n\to\infty}(1 + \frac{x}{n})^n = e^x$ and therefore

$$\lim_{\substack{n\to\infty\\ np=\lambda}} \varphi_{n,p}(t) = e^{\lambda(e^{i\hbar\omega t} - 1) + i\frac{\hbar\omega}{2} t}. \tag{A88}$$

If these are equal for all values of t, we deduce that up to arbitrary phases,

$$\lim_{\substack{n\to\infty\\ np=\lambda}} |n, p\rangle = |\sqrt{\lambda}\rangle \tag{A89}$$

where $|\sqrt{\lambda}\rangle$ is a coherent state.

These results are enough to prove convergence of the binomial state fluctuation relation to the coherent state fluctuation relation. The quantum distortion factors can also be obtained by perturbative means or by using the relevant quantities in the coherent state limit.

References

1. Guggenheim, E.A. The Thermodynamics of Magnetization. *Proc. R. Soc. Lond. Ser. A Math. Phys. Sci.* **1936**, *155*, 70–101. [CrossRef]
2. Alloul, H. *Thermodynamics of Superconductors*; Springer: Berlin/Heidelberg, Germany, 2011; pp. 175–199._6. [CrossRef]
3. Page, D.N. Hawking radiation and black hole thermodynamics. *New J. Phys.* **2005**, *7*, 203–203. [CrossRef]
4. Schrödinger, E.; Penrose, R. *What is Life?: With Mind and Matter and Autobiographical Sketches*; Canto, Cambridge University Press: Cambridge, UK, 1992; doi:10.1017/CBO9781139644129. [CrossRef]
5. Haynie, D. *Biological Thermodynamics*; Cambridge University Press: Cambridge, UK, 2001.
6. Ott, J.; Boerio-Goates, J. *Chemical Thermodynamics: Principles and Applications: Principles and Applications*; Elsevier Science: Amsterdam, The Netherlands, 2000.
7. Callen, H.; Callen, H.; Sons, W.. *Thermodynamics and an Introduction to Thermostatistics*; Wiley: Hoboken, NJ, USA, 1985.

8. Jarzynski, C. Equalities and Inequalities: Irreversibility and the Second Law of Thermodynamics at the Nanoscale. *Annu. Rev. Condens. Matter Phys.* **2011**, *2*, 329–351. [CrossRef]
9. Crooks, G.E. Entropy production fluctuation theorem and the nonequilibrium work relation for free energy differences. *Phys. Rev. E* **1999**, *60*, 2721–2726. [CrossRef]
10. Talkner, P.; Hänggi, P. The Tasaki–Crooks quantum fluctuation theorem. *J. Phys. Math. Theor.* **2007**, *40*, F569. [CrossRef]
11. Evans, D.J.; Searles, D.J. Equilibrium microstates which generate second law violating steady states. *Phys. Rev. E* **1994**, *50*, 1645–1648. [CrossRef]
12. Jarzynski, C. Nonequilibrium Equality for Free Energy Differences. *Phys. Rev. Lett.* **1997**, *78*, 2690–2693. [CrossRef]
13. Tasaki, H. Jarzynski Relations for Quantum Systems and Some Applications. *arXiv* **2000**, arXiv:cond-mat/0009244.
14. Kurchan, J. A Quantum Fluctuation Theorem. *arXiv* **2000**, arXiv:cond-mat/0007360.
15. Hänggi, P.; Talkner, P. The other QFT. *Nat. Phys.* **2015**, *11*, 108. [CrossRef]
16. Esposito, M.; Harbola, U.; Mukamel, S. Nonequilibrium fluctuations, fluctuation theorems, and counting statistics in quantum systems. *Rev. Mod. Phys.* **2009**, *81*, 1665–1702. [CrossRef]
17. Campisi, M.; Hänggi, P.; Talkner, P. Colloquium: Quantum fluctuation relations: Foundations and applications. *Rev. Mod. Phys.* **2011**, *83*, 771–791. [CrossRef]
18. Albash, T.; Lidar, D.A.; Marvian, M.; Zanardi, P. Fluctuation theorems for quantum processes. *Phys. Rev. E* **2013**, *88*, 032146. [CrossRef]
19. Manzano, G.; Horowitz, J.M.; Parrondo, J.M.R. Nonequilibrium potential and fluctuation theorems for quantum maps. *Phys. Rev. E* **2015**, *92*, 032129. [CrossRef]
20. Rastegin, A.E. Non-equilibrium equalities with unital quantum channels. *J. Stat. Mech. Theory Exp.* **2013**, *2013*, P06016. [CrossRef]
21. Campisi, M.; Talkner, P.; Hänggi, P. Fluctuation Theorem for Arbitrary Open Quantum Systems. *Phys. Rev. Lett.* **2009**, *102*, 210401. [CrossRef]
22. Jarzynski, C. Nonequilibrium work theorem for a system strongly coupled to a thermal environment. *J. Stat. Mech. Theory Exp.* **2004**, *2004*, P09005. [CrossRef]
23. Solinas, P.; Miller, H.J.D.; Anders, J. Measurement-dependent corrections to work distributions arising from quantum coherences. *Phys. Rev. A* **2017**, *96*, 052115. [CrossRef]
24. Allahverdyan, A.E. Nonequilibrium quantum fluctuations of work. *Phys. Rev. E* **2014**, *90*, 032137. [CrossRef] [PubMed]
25. Miller, H.J.D.; Anders, J. Time-reversal symmetric work distributions for closed quantum dynamics in the histories framework. *New J. Phys.* **2017**, *19*, 062001. [CrossRef]
26. Elouard, C.; Mohammady, M.H. Work, Heat and Entropy Production Along Quantum Trajectories. In *Thermodynamics in the Quantum Regime: Fundamental Aspects and New Directions*; Binder, F., Correa, L.A., Gogolin, C., Anders, J., Adesso, G., Eds.; Springer International Publishing: Cham, Switzerland, 2018; pp. 363–393._15. [CrossRef]
27. Horowitz, J.M. Quantum-trajectory approach to the stochastic thermodynamics of a forced harmonic oscillator. *Phys. Rev. E* **2012**, *85*, 31110. [CrossRef] [PubMed]
28. Elouard, C.; Herrera-Martí, D.A.; Clusel, M.; Auffèves, A. The role of quantum measurement in stochastic thermodynamics. *npj Quantum Inf.* **2017**, *3*, 9. [CrossRef]
29. Åberg, J. Fully Quantum Fluctuation Theorems. *Phys. Rev. X* **2018**, *8*, 11019. [CrossRef]
30. Alhambra, A.M.; Masanes, L.; Oppenheim, J.; Perry, C. Fluctuating Work: From Quantum Thermodynamical Identities to a Second Law Equality. *Phys. Rev. X* **2016**, *6*, 041017. [CrossRef]
31. Kwon, H.; Kim, M.S. Fluctuation Theorems for a Quantum Channel. *Phys. Rev. X* **2019**, *9*, 031029. [CrossRef]
32. Holmes, Z., The Coherent Crooks Equality. In *Thermodynamics in the Quantum Regime: Fundamental Aspects and New Directions*; Binder, F., Correa, L.A., Gogolin, C., Anders, J., Adesso, G., Eds.; Springer International Publishing: Cham, Switzerland, 2018; pp. 301–316._12. [CrossRef]
33. Horodecki, M.; Oppenheim, J. Fundamental limitations for quantum and nanoscale thermodynamics. *Nat. Commun.* **2013**, *4*, 2059. [CrossRef]
34. Brandão, F.; Horodecki, M.; Ng, N.; Oppenheim, J.; Wehner, S. The second laws of quantum thermodynamics. *Proc. Natl. Acad. Sci. USA* **2015**, *112*, 3275–3279. [CrossRef]

35. Åberg, J. Truly work-like work extraction via a single-shot analysis. *Nat. Commun.* **2013**, *4*, 1925. [CrossRef]
36. Gour, G.; Jennings, D.; Buscemi, F.; Duan, R.; Marvian, I. Quantum majorization and a complete set of entropic conditions for quantum thermodynamics. *Nat. Commun.* **2018**, *9*, 5352. [CrossRef]
37. Lostaglio, M.; Jennings, D.; Rudolph, T. Description of quantum coherence in thermodynamic processes requires constraints beyond free energy. *Nat. Commun.* **2015**, *6*, 6383. [CrossRef]
38. Åberg, J. Catalytic Coherence. *Phys. Rev. Lett.* **2014**, *113*, 150402. [CrossRef] [PubMed]
39. Korzekwa, K.; Lostaglio, M.; Oppenheim, J.; Jennings, D. The extraction of work from quantum coherence. *New J. Phys.* **2016**, *18*, 023045. [CrossRef]
40. Holmes, Z.; Weidt, S.; Jennings, D.; Anders, J.; Mintert, F. Coherent fluctuation relations: from the abstract to the concrete. *Quantum* **2019**, *3*, 124. [CrossRef]
41. Mingo, E.H.; Jennings, D. Decomposable coherence and quantum fluctuation relations. *Quantum* **2019**, *3*, 202. [CrossRef]
42. Zavatta, A.; Parigi, V.; Kim, M.S.; Bellini, M. Subtracting photons from arbitrary light fields: experimental test of coherent state invariance by single-photon annihilation. *New J. Phys.* **2008**, *10*, 123006. [CrossRef]
43. Ueda, M.; Imoto, N.; Ogawa, T. Quantum theory for continuous photodetection processes. *Phys. Rev. A* **1990**, *41*, 3891–3904. [CrossRef]
44. Barnett, S.M.; Ferenczi, G.; Gilson, C.R.; Speirits, F.C. Statistics of photon-subtracted and photon-added states. *Phys. Rev. A* **2018**, *98*, 013809. [CrossRef]
45. Zavatta, A.; Parigi, V.; Bellini, M. Experimental nonclassicality of single-photon-added thermal light states. *Phys. Rev. A* **2007**, *75*, 052106. [CrossRef]
46. Hloušek, J.; Ježek, M.; Filip, R. Work and information from thermal states after subtraction of energy quanta. *Sci. Rep.* **2017**, *7*, 13046. [CrossRef]
47. Vidrighin, M.D.; Dahlsten, O.; Barbieri, M.; Kim, M.S.; Vedral, V.; Walmsley, I.A. Photonic Maxwell's Demon. *Phys. Rev. Lett.* **2016**, *116*, 050401. [CrossRef]
48. Quantum communication with photon-added coherent states. *Quantum Inf. Process.* **2013**, *12*, 537–547. [CrossRef]
49. Braun, D.; Jian, P.; Pinel, O.; Treps, N. Precision measurements with photon-subtracted or photon-added Gaussian states. *Phys. Rev. A* **2014**, *90*, 013821. [CrossRef]
50. Mari, A.; Eisert, J. Positive Wigner Functions Render Classical Simulation of Quantum Computation Efficient. *Phys. Rev. Lett.* **2012**, *109*, 230503. [CrossRef]
51. Walschaers, M.; Fabre, C.; Parigi, V.; Treps, N. Entanglement and Wigner Function Negativity of Multimode Non-Gaussian States. *Phys. Rev. Lett.* **2017**, *119*, 183601. [CrossRef] [PubMed]
52. Perelomov, A. *Generalized Coherent States and Their Applications*; Theoretical and Mathematical Physics; Springer: Berlin/Heidelberg, Germany, 2012.
53. Arecchi, F.T.; Courtens, E.; Gilmore, R.; Thomas, H. Atomic Coherent States in Quantum Optics. *Phys. Rev. A* **1972**, *6*, 2211–2237. [CrossRef]
54. Stoler, D.; Saleh, B.; Teich, M. Binomial States of the Quantized Radiation Field. *Opt. Acta Int. J. Opt.* **1985**, *32*, 345–355. [CrossRef]
55. Vidiella-Barranco, A.; Roversi, J.A. Statistical and phase properties of the binomial states of the electromagnetic field. *Phys. Rev. A* **1994**, *50*, 5233–5241. [CrossRef]
56. Maleki, Y.; Maleki, A. Entangled multimode spin coherent states of trapped ions. *J. Opt. Soc. Am. B* **2018**, *35*, 1211–1217. [CrossRef]
57. Miry, S.R.; Tavassoly, M.K.; Roknizadeh, R. On the generation of number states, their single- and two-mode superpositions, and two-mode binomial state in a cavity. *J. Opt. Soc. Am. B* **2014**, *31*, 270–276. [CrossRef]
58. Garling, D.J.H. *Inequalities: A Journey into Linear Analysis*; Cambridge University Press: Cambridge, UK, 2007; doi:10.1017/CBO9780511755217. [CrossRef]
59. Baştuğ, T.; Kuyucak, S. Application of Jarzynski's equality in simple versus complex systems. *Chem. Phys. Lett.* **2007**, *436*, 383–387. [CrossRef]
60. West, D.K.; Olmsted, P.D.; Paci, E. Free energy for protein folding from nonequilibrium simulations using the Jarzynski equality. *J. Chem. Phys.* **2006**, *125*, 204910. [CrossRef] [PubMed]
61. Gittes, F. Two famous results of Einstein derived from the Jarzynski equality. *Am. J. Phys.* **2018**, *86*, 31–35. [CrossRef]

62. Deffner, S.; Jarzynski, C. Information Processing and the Second Law of Thermodynamics: An Inclusive, Hamiltonian Approach. *Phys. Rev. X* **2013**, *3*, 041003. [CrossRef]
63. Ballentine, L. *Quantum Mechanics: A Modern Development*; World Scientific: Singapore, 1998.
64. Crooks, G.E. Quantum operation time reversal. *Phys. Rev. A* **2008**, *77*, 034101. [CrossRef]
65. Jones, G.N.; Haight, J.; Lee, C.T. Nonclassical effects in the photon-added thermal state. *Quantum Semiclassical Opt. J. Eur. Soc. Part B* **1997**, *9*, 411–418. [CrossRef]
66. Usha Devi, A.R.; Prabhu, R.; Uma, M.S. Non-classicality of photon added coherent and thermal radiations. *Eur. Phys. J. D* **2006**, *40*, 133–138. [CrossRef]
67. Hu, L.-Y.; Fan, H.-Y. Wigner function and density operator of the photon-subtracted squeezed thermal state. *Chin. Phys. B* **2009**, *18*, 4657–4661. [CrossRef]
68. Bogdanov, Y.I.; Katamadze, K.G.; Avosopiants, G.V.; Belinsky, L.V.; Bogdanova, N.A.; Kalinkin, A.A.; Kulik, S.P. Multiphoton subtracted thermal states: Description, preparation, and reconstruction. *Phys. Rev. A* **2017**, *96*, 063803. [CrossRef]
69. Li, J.; Gröblacher, S.; Zhu, S.Y.; Agarwal, G.S. Generation and detection of non-Gaussian phonon-added coherent states in optomechanical systems. *Phys. Rev. A* **2018**, *98*, 011801. [CrossRef]
70. Fu, H.C.; Sasaki, R. Negative binomial and multinomial states: Probability distributions and coherent states. *J. Math. Phys.* **1997**, *38*, 3968–3987. [CrossRef]
71. Lee Loh, Y.; Kim, M. Visualizing spin states using the spin coherent state representation. *Am. J. Phys.* **2015**, *83*, 30–35. [CrossRef]
72. Sperling, J.; Walmsley, I.A. Quasiprobability representation of quantum coherence. *Phys. Rev. A* **2018**, *97*, 062327. [CrossRef]
73. Atkins, P.W.; Dobson, J.C.; Coulson, C.A. Angular momentum coherent states. *Proc. R. Soc. London. Math. Physical Sci.* **1971**, *321*, 321–340. [CrossRef]
74. Zelaya, K.D.; Rosas-Ortiz, O. Optimized Binomial Quantum States of Complex Oscillators with Real Spectrum. *J. Phys. Conf. Ser.* **2016**, *698*, 12026. [CrossRef]
75. Skotiniotis, M.; Gour, G. Alignment of reference frames and an operational interpretation for the G-asymmetry. *New J. Phys.* **2012**, *14*, 073022. [CrossRef]
76. Peruzzo, A.; McClean, J.; Shadbolt, P.; Yung, M.H.; Zhou, X.Q.; Love, P.J.; Aspuru-Guzik, A.; O'Brien, J.L. A variational eigenvalue solver on a photonic quantum processor. *Nat. Commun.* **2014**, *5*, 4213. [CrossRef]
77. Kieferová, M.; Scherer, A.; Berry, D.W. Simulating the dynamics of time-dependent Hamiltonians with a truncated Dyson series. *Phys. Rev. A* **2019**, *99*, 042314. [CrossRef]
78. Lukacs, E. *Characteristic Functions*; Griffin Books of Cognate Interest; Hafner Publishing Company: New York, NY, USA, 1970.

 © 2020 by the authors. Licensee MDPI, Basel, Switzerland. This article is an open access article distributed under the terms and conditions of the Creative Commons Attribution (CC BY) license (http://creativecommons.org/licenses/by/4.0/).

Article

Joint Fluctuation Theorems for Sequential Heat Exchange

Jader Santos [1], André Timpanaro [2] and Gabriel Landi [1,*]

[1] Instituto de Física da Universidade de São Paulo, São Paulo 05314-970, Brazil; jader.pereira.santos@gmail.com
[2] Universidade Federal do ABC, Santo André 09210-580, Brazi; a.timpanaro@ufabc.edu.br
* Correspondence: gtlandi@if.usp.br

Received: 5 March 2020; Accepted: 8 July 2020; Published: 12 July 2020

Abstract: We study the statistics of heat exchange of a quantum system that collides sequentially with an arbitrary number of ancillas. This can describe, for instance, an accelerated particle going through a bubble chamber. Unlike other approaches in the literature, our focus is on the *joint* probability distribution that heat Q_1 is exchanged with ancilla 1, heat Q_2 is exchanged with ancilla 2, and so on. This allows us to address questions concerning the correlations between the collisional events. For instance, if in a given realization a large amount of heat is exchanged with the first ancilla, then there is a natural tendency for the second exchange to be smaller. The joint distribution is found to satisfy a Fluctuation theorem of the Jarzynski–Wójcik type. Rather surprisingly, this fluctuation theorem links the statistics of multiple collisions with that of independent single collisions, even though the heat exchanges are statistically correlated.

Keywords: fluctuation theorems; collisional models

1. Introduction

Fluctuations of thermodynamic quantities, which are usually negligible in macroscopic systems, are known to play a dominant role in the micro- and mesoscopic domain. These fluctuations are embodied in the so-called fluctuation theorems (FT) [1–4], a collection of predictions for systems evolving under nonequilibrium conditions valid beyond linear response. They can be summarized as [5,6]

$$\frac{P(+\Sigma)}{\tilde{P}(-\Sigma)} = e^{\Sigma}, \tag{1}$$

where $P(\Sigma)$ denotes the probability that an amount of entropy Σ is produced in a certain process and $\tilde{P}(\Sigma)$ denotes the corresponding probability for the time-reversed process.

Of the many scenarios which present FTs, one which is particularly interesting is that of heat exchange between a system S, prepared in equilibrium with a temperature T_s, and an environment E, prepared in a different temperature T_e. In this case, as first shown by Jarzynski and Wójcik in Ref. [7], the distribution $P(Q)$ of the heat exchanged between them, satisfies

$$\frac{P(+Q)}{\tilde{P}(-Q)} = e^{\Delta \beta Q}, \tag{2}$$

where $\Delta\beta = \beta_e - \beta_s$ (with $\beta = 1/T$ and $k_B = 1$). Here, and throughout the paper, Q denotes the net heat transfer from the system to the environment. Quite surprising, in this case it turns out that $\tilde{P}(Q) = P(Q)$, meaning the statistics of the forward and backward processes are the same. Equation (2) was subsequently generalized to allow for the exchange of both energy and particles between several interacting systems initially at different temperatures and chemical potentials [6,8,9].

Here we consider a generalization of this scenario, where the system interacts sequentially with multiple parts of the environment, exchanging heat with each part. One can imagine, for instance, an accelerated particle crossing a bubble chamber. In this case, the system will leave a trail on E, represented by the heat exchanged in each point. In the microscopic domain this process will be stochastic, with a random amount of heat exchanged in each interaction.

The key idea that we will explore in this paper is to look at the joint probability distribution for the heat exchanged with each part, $P(Q_1, Q_2, Q_3, \ldots)$. This allows us to understand the correlations between the different heat exchanges.

For instance, in a situation where all the ancillas have the same temperature, from a stochastic perspective a large exchange in the first collision increases the probability that the second collision exchanges less. This feature is fully captured by the joint distribution. This happens because thermal operations have the property of bringing the system closer to its thermal equilibrium state, σ_{eq}, i.e., [10]

$$D(\sigma_0 \| \sigma_{\text{eq}}) \geq D(\sigma_1 \| \sigma_{\text{eq}}) \geq D(\sigma_2 \| \sigma_{\text{eq}}) \geq \cdots \geq D(\sigma_N \| \sigma_{\text{eq}}) \tag{3}$$

where $D(\rho' \| \rho) = \text{Tr}(\rho' \ln \rho' - \rho' \ln \rho)$ is the quantum relative entropy. If in the first interaction the system exchange a large quantity of heat, the system gets a lot closer to its steady state. So in the next interaction, the system should exchange less heat.

To formalize this idea, we split the environment into a set of ancillas A_i, with which the system interacts sequentially, producing a collisional model [11–14]. The process is schematically illustrated in Figure 1 and the formal framework is developed in Section 2. In Section 3 we then show that $P(Q_1, Q_2, Q_3, \ldots)$ satisfies a fluctuation theorem that generalizes (2). Moreover, we show how this fluctuation theorem relates the joint distribution to the statistics of a single collision, even though the events are statistically correlated.

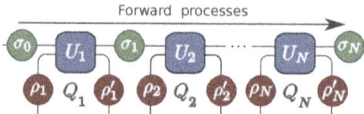

Figure 1. Schematic representation of a system S interacting sequentially with a series of ancillas. The system starts in the state σ_0 and the ancillas in an initial states ρ_i, which are assumed to be thermal but at possibly different temperatures. Each SA_i interaction is also governed by a possibly different unitary U_i.

2. Formal Framework

We consider a quantum system S, with Hamiltonian H^s, prepared in a thermal state $\sigma_0 = e^{-\beta_s H^s}/Z_s$, with temperature T_s. The system is put to interact sequentially with a series of N ancillas A_i, as depicted in Figure 1. The ancillas are not necessarily identical. Each has Hamiltonian H^i and is prepared in a thermal state $\rho_i = e^{-\beta_i H^i}/Z_i$, with possibly different temperatures T_i. Each collision is described by a unitary operator U_i acting only on SA_i, which may also differ from one interaction to another.

In order to comply with the scenario of Ref. [7], we assume that the U_i satisfy the strong energy-preservation condition

$$[U_i, H^s + H^i] = 0. \tag{4}$$

Or, what is equivalent, that each collision is a thermal operation [10,15]. This implies that all energy that leaves S enters A_i, so nothing is stuck in the interaction. As a consequence, there is no work involved and all the change in energy of the system can be unambiguously identified as heat flowing to the ancillas [13].

We label the eigenvalues and eigenvectors of the system as $H^s|\alpha\rangle = E^s_\alpha|\alpha\rangle$. For concreteness, we assume these levels are non-degenerate. Time is labeled discretely by $i = 1, 2, 3, \ldots$, representing which collisions already took place. For instance, the initial state is decomposed as $\sigma_0 = \sum_{\alpha_0} p_0(\alpha_0)|\alpha_0\rangle\langle\alpha_0|$, with $p_0(\alpha_0) = e^{-\beta_s E^s_{\alpha_0}}/Z_s$ and we use α_0 to emphasize that this is before the first collision. Similarly, the eigenvalues and eigenvectors of the ancillas are labeled as $H_i|n_i\rangle = E^e_{n_i}|n_i\rangle$. The initial state of each A_i is thus decomposed as $\rho_i = \sum_{n_i} q_i(n_i)|n_i\rangle\langle n_i|$ where $q_i(n_i) = e^{-\beta_i E^e_{n_i}}/Z_i$.

The dynamics depicted in Figure 1 generates a stroboscopic map for the system. The joint state of SA_i after the interaction is given by

$$\varrho_i = U_i(\sigma_{i-1} \otimes \rho_i)U_i^\dagger. \tag{5}$$

Taking the partial trace over A_i then leads to the updated state σ_i. Conversely, tracing over the system leads to the reduced state ρ'_i of the ancilla after the interaction (Figure 1).

The fact that the unitary is energy preserving (Equation (4)), together with the assumption that the energy levels are non-degenerate, means that it is possible to construct quantum trajectories for the system in two equivalent ways. The first is to assume a two-point measurement scheme in S at each step [16,17]. Equation (4) implies that the system will remain diagonal in the energy basis, so that measurements in this basis are non-invasive (that is, have no additional entropy production associated to it). Measuring S in the energy basis after each collision then leads to the trajectory

$$\gamma_s = \{\alpha_0, \alpha_1, \ldots, \alpha_N\}. \tag{6}$$

The heat associated with each collision is then readily defined as

$$Q_i[\gamma_s] = -E^s_{\alpha_i} + E^s_{\alpha_{i-1}}, \tag{7}$$

Alternatively, one can construct a quantum trajectory by measuring the ancillas, before and after each collision, plus a single measurement of the system before the process starts. That is, one can consider instead a quantum trajectory of the form

$$\gamma_e = \{\alpha_0, n_1, n'_1, n_2, n'_2, \ldots, n_N, n'_N\}. \tag{8}$$

This, in a sense, is much more natural since the ancillas are only used once and thus may be experimentally more easily accessible. Furthermore, as far as heat exchange is concerned, this turns out to be equivalent to the trajectory (6). The reason is that Equation (4) implies the restriction

$$\langle \alpha_i n'_i | U_i | \alpha_{i-1} n_i \rangle \propto \delta((E^s_{\alpha_i} + E^e_{n'_i}) - (E^s_{\alpha_{i-1}} + E^e_{n_i})) \tag{9}$$

where $\delta(x)$ is the Kronecker delta. In addition, since the energy values are taken to be non-degenerate, energies uniquely label states. Thus, for instance, if we know α_0, n_1, n'_1 we can uniquely determine α_1, and so on. The converse, however, is not true: from α_0 and α_1 we cannot specify n_1 and n'_1 (which is somewhat evident given that the number of points in Equation (6) is smaller than that in Equation (8)). This, however, is not a problem if one is interested only in the heat exchanged, which can also be defined from the trajectory (8) as

$$Q_i[\gamma_e] = E^e_{n'_i} - E^e_{n_i}. \tag{10}$$

Due to Equation (9), this must coincide with Equation (7); i.e., $Q_i[\gamma_e] \equiv Q_i[\gamma_s]$.

The assumption in Equation (4) may at first seem somewhat artificial. However, this is not the case. This assumption is a way to bypass the idea of weak coupling, which is one of the conditions used in [7]. Moreover, the interesting thing about the present analysis is that it establishes under which conditions Equations (6) and (8) are equivalent. Naively one would expect that this is often the case. However, as the above arguments show, several assumptions are necessary for this to be the case. This reflects some of the challenges that appear in describing thermodynamics in the quantum regime.

2.1. Path Probabilities from Measurements in S

Thermal operations imply that the probability that, after the i-th collision, the system is in a given eigenstate $|\alpha_i\rangle$ depends only on the probabilities in the previous time. That is, the dynamics of populations and coherences completely decouple [18]. Indeed, Equation (5) together with Equation (4) imply that

$$p_i(\alpha_i) = \langle \alpha_i | \sigma_i | \alpha_i \rangle = \sum_{\alpha_{i-1}} M_i(\alpha_i | \alpha_{i-1}) p_{i-1}(\alpha_{i-1}), \tag{11}$$

where

$$M_i(\alpha_i | \alpha_{i-1}) = \sum_{n_i, n_i'} |\langle \alpha_i, n_i' | U_i | \alpha_{i-1}, n_i \rangle|^2 q_i(n_i). \tag{12}$$

The populations therefore evolve as a classical Markov chain, with $M_i(\alpha_i|\alpha_{i-1})$ representing the transition probability of going from α_{i-1} to α_i. Moreover, Equation (9) together with the fact that the ancillas are initially thermal, imply that $M_i(\alpha_i|\alpha_{i-1})$ satisfies detailed balance

$$M_i(\alpha_i|\alpha_{i-1}) e^{-\beta_i E^s_{\alpha_{i-1}}} = M_i(\alpha_{i-1}|\alpha_i) e^{-\beta_i E^s_{\alpha_i}}, \tag{13}$$

where, notice, what appears here is the temperature β_i of ancilla A_i.

The path probability associated with γ_s in Equation (6) will then be

$$\mathcal{P}[\gamma_s] = M_N(\alpha_N|\alpha_{N-1}) \ldots M_2(\alpha_2|\alpha_1) M_1(\alpha_1|\alpha_0) p_0(\alpha_0), \tag{14}$$

which is nothing but the joint distribution of a Markov chain. We call attention to the clear causal structure of this expression: marginalizing over future events has no influence on past ones. For instance, summing over α_N leads to a distribution of the exact same form. Conversely, marginalizing over past variables completely changes the distribution.

The joint distribution of heat can then be constructed from Equation (14) in the usual way:

$$P(Q_1, \ldots, Q_N) = \sum_{\gamma_s} \mathcal{P}[\gamma_s] \left(\prod_{i=1}^{N} \delta(Q_i - Q_i[\gamma_s]) \right). \tag{15}$$

This is the basic object that we will explore in this paper.

2.2. Path Probabilities from Measurements in the A_i

Alternatively, we also wish to show how Equation (15) can be constructed from the trajectory γ_e in Equation (8). The easiest way to accomplish this is to first consider the augmented trajectory

$$\gamma_{se} = \{\alpha_0, n_1, n_1', \alpha_1, n_2, n_2', \alpha_2, \ldots, n_N, n_N', \alpha_N\} \tag{16}$$

Introducing the transition probabilities $R_i(\alpha_i, n_i'|\alpha_{i-1}, n_i) = |\langle \alpha_i, n_i'|U_i|\alpha_{i-1}, n_i\rangle|^2$, the path distribution associated with the augmented trajectory γ_{se} will be

$$\mathcal{P}[\gamma_{se}] = R_N(\alpha_N, n_N'|\alpha_{N-1}, n_N) \ldots R_1(\alpha_1, n_1'|\alpha_0, n_1) q_N(n_N) \ldots q_1(n_1) p_0(\alpha_0).$$

As a sanity check, if we marginalize this over n_i and n_i' we find

$$P[\gamma_s] = \sum_{\substack{n_1,\ldots,n_N \\ n'_1,\ldots,n'_N}} R_N(\alpha_N, n'_N|\alpha_{N-1}, n_N)\ldots R_1(\alpha_1, n'_1|\alpha_0, n_1) q_N(n_N)\ldots q_1(n_1) p_0(\alpha_0)$$

$$= M_N(\alpha_N|\alpha_{N-1})\ldots M_2(\alpha_2|\alpha_1) M_1(\alpha_1|\alpha_0) p_0(\alpha_0),$$

where we used Equation (12). This is therefore precisely $P[\gamma_s]$ in Equation (14), as expected. Instead, from $P[\gamma_{se}]$ one can now obtain $P[\gamma_e]$ by marginalizing over $\alpha_1,\ldots,\alpha_N$; viz.,

$$P[\gamma_e] = \sum_{\alpha_1,\ldots,\alpha_N} R_N(\alpha_N, n'_N|\alpha_{N-1}, n_N)\ldots R_1(\alpha_1, n'_1|\alpha_0, n_1) q_N(n_N)\ldots q_1(n_1) p_0(\alpha_0). \qquad (17)$$

The above analysis puts in evidence the Hidden Markov nature of the dynamics in Figure 1. When measurements are done in the ancilla, the system plays the role of the hidden layer, which is not directly accessible. Instead, predictions about the system must be made from the visible layer (i.e., the ancillas).

This Hidden Markov nature manifests itself on the fact that even though the system obeys a Markov chain [Equation (14)], the same is not true for the ancillas. In symbols, this is manifested by the fact that n'_i depends not only on n_i and n'_{i-1}, but on the entire past history $(n_1, n'_1, \ldots, n_{i-1}, n'_{i-1}, n_i)$. This is intuitive in a certain sense: the amount of heat exchanged at the i-th collision will depend on the heat exchanged in all past events.

With $P[\gamma_e]$, the distribution of heat, Equation (15) can be equivalently defined using Equation (10). One then finds

$$P(Q_1,\ldots,Q_N) = \sum_{\gamma_e} P[\gamma_e] \left(\prod_{i=1}^{N} \delta(Q_i - Q_i[\gamma_e]) \right). \qquad (18)$$

The reason why this is equivalent to Equation (15) becomes clear from the way we derived $P[\gamma_e]$ above: we can expand the summation to γ_{se} and then use the fact that $Q_i[\gamma_s] = Q_i[\gamma_e]$.

2.3. Backward Process

To construct the fluctuation theorem, we must now establish the backward process. As shown in [19], however, there is an arbitrariness in the choice of the initial state of the backward process; different choices lead to different definitions of the entropy production. Here we are interested specifically in heat and the generalization of the Jarzynski–Wójcik fluctuation theorem [7]. Hence, we assume that in the backward process both system and ancillas are fully reset back to their thermal states. As usual, the time-reversed interaction between SA_i now takes place by means of the unitary $U_i^\dagger$. However, the order of the interactions must now be flipped around, as shown in Figure 2. More about the choice of backward process can be found in [20,21] and its relation to the notion of recovery maps is discussed in [22].

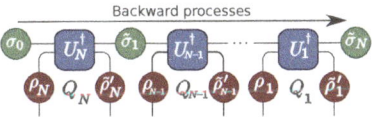

Figure 2. Schematic representation of the backward process.

In the backward process, the system will therefore evolve according to

$$\tilde{p}_i(\alpha_{N-i}) = \sum_{\alpha_{N-i+1}} M_{N-i+1}(\alpha_{N-i}|\alpha_{N-i+1}) \tilde{p}_{i-1}(\alpha_{N-i+1}),$$

where we index the states as α_{N-i} instead of α_i just so that the trajectory γ_s can remain the same as in the forward process. The path probability $\tilde{\mathcal{P}}[\gamma_s]$ associated to this process will then be

$$\tilde{\mathcal{P}}[\gamma_s] = M_1(\alpha_0|\alpha_1)\ldots M_N(\alpha_{N-1}|\alpha_N)p_0(\alpha_N), \tag{19}$$

which is similar to that used in the original Crooks fluctuation theorem [23]. The corresponding heat distribution is

$$\tilde{P}(Q_N,\ldots,Q_1) = \sum_{\gamma_s} \tilde{\mathcal{P}}[\gamma_s] \prod_{i=1}^{N} \delta(Q_i + Q_i[\gamma_s]), \tag{20}$$

where Q_i continues to be the heat exchanged with A_i (which is now different from the heat exchanged at collision i).

3. Joint Fluctuation Theorem for Heat Exchange

We are now ready to construct the fluctuation theorem. The detailed balance condition (13) immediately implies that Equations (15) and (20) will be related by

$$\frac{P(Q_1,\ldots,Q_N)}{\tilde{P}(-Q_N,\ldots,-Q_1)} = e^{\sum_{i=1}^{N}(\beta_i - \beta_s)Q_i}. \tag{21}$$

This is a theorem for the joint distribution of the heat exchanged between multiple ancillas. It thus represents a generalization of Ref. [7] to the case where the system interacts sequentially with multiple reservoirs. This result has several features which are noteworthy. First, note that the temperature β_i of the ancillas are not necessarily the same. Second, note how after the first collision the state of the system is no longer thermal. However, still, this does not affect the fluctuation theorem. All that matters is that before the first collision the system is in equilibrium.

It is also important to point out that any Markov chain satisfying the detailed balance relation also satisfies a fluctuation theorem [24]. This fact can be used to obtain Equation (21) when properly choosing the rates of the Markovian evolution. Beyond that, a generalization of the detailed FT to multiple reservoirs has also being obtained before, e.g., in Ref. [25].

3.1. Causal Order and Relation to Single Collisions

The causal order of the process plays a crucial role here. Marginalizing over future events has no effect on the fluctuation theorem. That is, from (21) one could very well construct a similar relation for $P(Q_1,\ldots,Q_{N-1})$, by simply summing over Q_N. This is not possible, however, for marginalization over past events. That is, $P(Q_2,\ldots,Q_N)$, for instance, does not satisfy a fluctuation theorem.

The right-hand side of Equation (21) is very similar to what appears in the original FT (2). We can make this more rigorous as follows. Let us consider a different process, consisting of a single collision between the system thermalized in β_s and an ancilla thermalized in β_i (Figure 3). The associated heat distribution $P_{sc}(Q_i)$ will then satisfy Equation (2); viz.,

$$\frac{P_{sc}(Q_i)}{P_{sc}(-Q_i)} = e^{(\beta_i - \beta_s)Q_i}, \tag{22}$$

where, recall, in this case of a single collision the backward process coincides with the forward one, so that the distribution $\tilde{P}_{sc}$ in the denominator is simply P_{sc}. It is very important to emphasize, however, that $P_{sc}(Q_i)$ is not the marginal of $P(Q_1,\ldots,Q_N)$ (with the exception of Q_1). Notwithstanding, comparing with Equation (21), we see that the full process in Figure 1 is related to the single-collision processes according to

$$\frac{P(Q_1,\ldots,Q_N)}{\tilde{P}(-Q_N,\ldots,-Q_1)} = \frac{P_{sc}(Q_1)}{P_{sc}(-Q_1)} \cdots \frac{P_{sc}(Q_N)}{P_{sc}(-Q_N)}. \tag{23}$$

This result is noteworthy, for the right-hand side is a product whereas the left-hand side is not. The full distribution $P(Q_1,\ldots,Q_N)$ cannot be expressed as a product because the heat exchanges are, in general, not statistically independent. Notwithstanding, the ratio on the left-hand side of (23) does factor into a product. The point, though, is that this is not the product of the marginals, but of another distribution P_{sc}.

One can also write a formula of the form (23), but for only some of the heat exchanges. For instance, it is true that

$$\frac{P(Q_1,\ldots,Q_N)}{\tilde{P}(-Q_N,\ldots,-Q_1)} = \frac{P(Q_1,\ldots,Q_{N-1})}{\tilde{P}(-Q_{N-1},\ldots,-Q_1)} \frac{P_{sc}(Q_N)}{P_{sc}(-Q_N)}. \qquad (24)$$

This kind of decomposition, however, depends crucially on the causal structure since it can only be done for future exchanges. For instance, we cannot write something involving $P(Q_2,\ldots,Q_N)$. The reason is that $P(Q_1,\ldots,Q_{N-1})$ satisfies the fluctuation theorem (21), but $P(Q_2,\ldots,Q_N)$ does not (since, after the first collision the system is no longer in a thermal state).

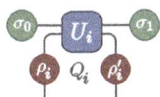

Figure 3. Schematic representation of a single collision event.

3.2. Information-Theoretic Formulation of the Entropy Production

We define the entropy production associated with Equation (21) as

$$\Sigma[\gamma_s] = \ln \frac{P[\gamma_s]}{\tilde{P}[\gamma_s]} = \sum_{i=1}^{N}(\beta_i - \beta_s)Q_i[\gamma_s]. \qquad (25)$$

The second equality is obtained using the detailed balance relation (13). We emphasize that this is the entropy production associated with the choice of backward protocol used in Section 2.3, which may differ from other definitions in the literature [18,26]. As discussed in [19], the interpretation of the entropy production depends on the choice of the initial state of the backwards process. For instance, if we have chosen the initial state as the final state of the forward process, i.e., the state ϱ_N (see Equation (5)), we would have a a contribution related to the correlations between the system and the ancillas. This type of entropy production was called the inclusive entropy production in Ref. [19]. This happens because this state carries the information about the correlations. Here we have choose a initial state for the backward process that does not have this contributions.

In [7], Jarzynski and Wójcik calculated an upper bound on the probability of observing a violation of the second law, i.e., the passage of heat from a colder to a hotter body. We can apply the same reasoning to Equation (25). Let us assume that all ancillas start in the same thermal state with temperature T_a and $\beta_a - \beta_s > 0$. The probability that the heat transfer from the system to i-th ancilla will fall below a specified value q_i in each interaction through the whole process, obeys the inequality

$$\int_{-\infty}^{q_1} dQ_1 \cdots \int_{-\infty}^{q_N} dQ_N \; P(Q_1,\ldots,Q_N) \leq e^{(\beta_a-\beta_s)(q_1+\cdots+q_N)} \qquad (26)$$

which is the multiple-exchange extension of the result obtained in [7]. Equation (26) shows that observing a positive total transference of heat from the hot system to the cold ancillas dies exponentially with $q_1 + \cdots + q_N$.

Alternatively, we can consider the entropy production from the perspective of the global trajectory γ_{se} in Equation (16). Using also that $Q_i[\gamma_s] = Q_i[\gamma_e]$, we can then write $\Sigma[\gamma_{se}]$ as

$$\Sigma[\gamma_{se}] = \sum_{i=1}^{N} \beta_i Q_i[\gamma_e] - \beta_s(E^s_{\alpha_N} - E^s_{\alpha_0}) = \sum_{i=1}^{N} \ln \frac{q_i(n_i)}{q_i(n'_i)} + \ln \frac{p_0(\alpha_0)}{p_0(\alpha_N)}. \tag{27}$$

The average entropy production may then be written as

$$\langle \Sigma[\gamma_{se}] \rangle = S(\sigma_N) - S(\sigma_0) + D(\sigma_N || \sigma_0) + \sum_{i=1}^{N} \left\{ S(\rho'_i) - S(\rho_i) + D(\rho'_i || \rho_i) \right\}, \tag{28}$$

where $S(\rho) = -\text{Tr}(\rho \ln \rho)$ is the von Neumann entropy. Here σ_N is the final state of the system after the N collisions. In the Equation (28), we can identify

$$S(\sigma_N) - S(\sigma_0) + \sum_{i=1}^{N} S(\rho'_i) - S(\rho_i) = \Delta I_{se} \tag{29}$$

where ΔI_{se} is the change in the mutual information between the system and the ancillas. This way we can have a more clear meaning of the expression (28). One term is proportional to the total correlations built between system and ancillas and the other two relative entropy terms measure the disturbance on the environment and the system during the process.

The important aspect of this result is that it depends only on local changes in the ancillas. That is, all quantities refer to the local states ρ'_i of each ancilla after the interaction. In reality, because the ancillas all interact with the system, they actually become indirectly correlated. These correlations are still represented indirectly in $\Sigma[\gamma_{se}]$, but they do not appear explicitly. This, ultimately, is a consequence of the choice of backward process that is used in the Jarzynski–Wójcik scenario [7].

3.3. Initially Correlated Ancillas

One possible extension of our formalism is to consider the case of initially correlated system-ancillas. In this case, we could explore how the correlation between the system and the ancillas affect the XFT. This problem was studied for a single heat exchange in [27] and in our case, the same approach yields

$$\frac{P[\gamma_{se}]}{\tilde{P}[\gamma_{se}]} = e^{-\Delta \mathcal{I}(\gamma_{se}) + \sum_{i=1}^{N}(\beta_s - \beta_i)(E^s_{\alpha_i} - E^s_{\alpha_{i-1}}) + \sum_{i=1}^{N} \beta_i [E^s_{\alpha_i} + E^e_{n'_i} - (E^s_{\alpha_{i-1}} + E^e_{n_i})]} \tag{30}$$

where the $\Delta \mathcal{I}(\gamma_{se}) = I^* - I$ with

$$I^* = \ln \left[\frac{p(\alpha_n, n'_1, \ldots, n'_N)}{p_0(\alpha_N) q_1(n'_1) \ldots q_N(n'_N)} \right] \tag{31}$$

$$I = \ln \left[\frac{p(\alpha_0, n_1, \ldots, n_N)}{p_0(\alpha_0) q_1(n_1) \ldots q_N(n_N)} \right] \tag{32}$$

where we define $p(\alpha_0, n_1, \ldots, n_N) = \langle \alpha_0, n_1, \ldots, n_N | \rho_{SE} | \alpha_0, n_1, \ldots, n_N \rangle$. Here ρ_{SE} is the initial state for the system-ancillas. This result is similar to the one found in [27]. Because in our case, we are working with thermal operations, we can write Equation (30) as

$$\frac{P[\gamma_{se}]}{\tilde{P}[\gamma_{se}]} = e^{-\Delta \mathcal{I}(\gamma_{se}) + \sum_{i=1}^{N}(\beta_s - \beta_i)(E^s_{\alpha_i} - E^s_{\alpha_{i-1}})} \tag{33}$$

By taking the above equation and sum over all trajectories, to obtain the nonequilibrium equality for an initially correlated state

$$\langle e^{\Delta \mathcal{I} + \sum_{i=1}^{N}(\beta_s - \beta_i) Q_i} \rangle = 1 \tag{34}$$

and then using Jensen's inequality we have that

$$\sum_{i=1}^{N}(\beta_i - \beta_s)\langle Q_i \rangle \geq \langle \Delta \mathcal{I} \rangle \qquad (35)$$

So it is possible to obtain a type of Clausius relation where now the entropy production has a new lower bound.

4. Conclusions

To summarize, we have considered here the sequential heat exchange between a system and a series of ancillas. We assume all entities start in thermal state, but at possibly different temperatures. Moreover, all interactions are assumed to be described by thermal operations, which makes the identification of heat unambiguous. The main object of our study was the joint probability of heat exchange $P(Q_1, \ldots, Q_N)$ for a set of N collisions. This object contemplates the correlations between heat exchange, a concept which to the best of our knowledge, has not been explored in the quantum thermodynamics community. We showed that $P(Q_1, \ldots, Q_N)$ satisfies a fluctuation theorem, which relates this joint distribution with single collision events. This result, we believe, could serve to highlight the interesting prospect of analyzing thermodynamic quantities in time-series and other sequential models.

Author Contributions: J.S., A.T. and G.L. equally contributed to conceptualization, investigation and writing of the paper. All authors have read and agreed to the published version of the manuscript.

Funding: G.L. acknowledges the financial support from the São Paulo funding agency FAPESP and from the Instituto Nacional de Ciência e Tecnologia em Informação Quântica. J.S. would like to acknowledge the financial support from the CAPES/PNPD (process No. 88882.315481/2013-01).

Acknowledgments: The authors acknowledge E. Lutz and M. Paternostro for fruitful discussions.

Conflicts of Interest: The authors declare no conflict of interest.

References

1. Evans, D.J.; Cohen, E.G.; Morriss, G.P. Probability of second law violations in shearing steady states. *Phys. Rev. Lett.* **1993**, *71*, 2401–2404. [CrossRef] [PubMed]
2. Gallavotti, G.; Cohen, E.G.D. Dynamical ensembles in nonequilibrium statistical mechanics. *Phys. Rev. Lett.* **1995**, *74*, 2694–2697. [CrossRef] [PubMed]
3. Jarzynski, C. Equilibrium free-energy differences from nonequilibrium measurements: A master-equation approach. *arXiv* **1997**, arXiv:cond-mat/9707325.
4. Crooks, G.E. Entropy production fluctuation theorem and the nonequilibrium work relation for free energy differences. *Phys. Rev. E* **1999**. [CrossRef]
5. Campisi, M.; Hänggi, P.; Talkner, P. Colloquium: Quantum fluctuation relations: Foundations and applications. *Rev. Mod. Phys.* **2011**, *83*, 771–791. [CrossRef]
6. Esposito, M.; Harbola, U.; Mukamel, S. Nonequilibrium fluctuations, fluctuation theorems, and counting statistics in quantum systems. *Rev. Mod. Phys.* **2009**, *81*, 1665–1702. [CrossRef]
7. Jarzynski, C.; Wójcik, D.K. Classical and Quantum Fluctuation Theorems for Heat Exchange. *Phys. Rev. Lett.* **2004**, *92*, 230602. [CrossRef]
8. Saito, K.; Utsumi, Y. Symmetry in full counting statistics, fluctuation theorem, and relations among nonlinear transport coefficients in the presence of a magnetic field. *Phys. Rev. Condens. Matter Mater. Phys.* **2008**, *78*, 1–7. [CrossRef]
9. Andrieux, D.; Gaspard, P. Fluctuation theorem for transport in mesoscopic systems. *J. Stat. Mech. Theory Exp.* **2006**. [CrossRef]
10. Brandão, F.G.S.L.; Horodecki, M.; Oppenheim, J.; Renes, J.M.; Spekkens, R.W. Resource Theory of Quantum States Out of Thermal Equilibrium. *Phys. Rev. Lett.* **2013**, *111*, 250404. [CrossRef]
11. Rodrigues, F.L.; De Chiara, G.; Paternostro, M.; Landi, G.T. Thermodynamics of Weakly Coherent Collisional Models. *Phys. Rev. Lett.* **2019**, *123*, 140601. [CrossRef] [PubMed]

12. Strasberg, P.; Schaller, G.; Brandes, T.; Esposito, M. Quantum and Information Thermodynamics: A Unifying Framework Based on Repeated Interactions. *Phys. Rev. X* **2017**. [CrossRef]
13. De Chiara, G.; Landi, G.; Hewgill, A.; Reid, B.; Ferraro, A.; Roncaglia, A.J.; Antezza, M. Reconciliation of quantum local master equations with thermodynamics. *New J. Phys.* **2018**, *20*, 113024. [CrossRef]
14. Scully, M.O.; Suhail Zubairy, M.; Agarwal, G.S.; Walther, H. Extracting work from a single heat bath via vanishing quantum coherence. *Science* **2003**, *299*, 862–864. [CrossRef]
15. Brandão, F.; Horodecki, M.; Ng, N.; Oppenheim, J.; Wehner, S. The second laws of quantum thermodynamics. *Proc. Natl. Acad. Sci. USA* **2015**, *112*, 3275–3279. [CrossRef]
16. Campisi, M.; Talkner, P.; Hänggi, P. Fluctuation theorems for continuously monitored quantum fluxes. *Phys. Rev. Lett.* **2010**, *105*, 1–4. [CrossRef]
17. Utsumi, Y.; Golubev, D.S.; Marthaler, M.; Saito, K.; Fujisawa, T.; Schön, G. Bidirectional single-electron counting and the fluctuation theorem. *Phys. Rev. Condens. Matter Mater. Phys.* **2010**, *81*, 1–5. [CrossRef]
18. Santos, J.P.; Céleri, L.C.; Landi, G.T.; Paternostro, M. The role of quantum coherence in non-equilibrium entropy production. *npj Quantum Inf.* **2019**, *5*, 23. [CrossRef]
19. Manzano, G.; Horowitz, J.M.; Parrondo, J.M.R. Quantum Fluctuation Theorems for Arbitrary Environments: Adiabatic and Nonadiabatic Entropy Production. *Phys. Rev.* **2018**, *8*, 031037. [CrossRef]
20. Alhambra, Á.M.; Woods, M.P. Dynamical maps, quantum detailed balance, and the Petz recovery map. *Phys. Rev.* **2017**, *96*, 022118. [CrossRef]
21. Åberg, J. Fully Quantum Fluctuation Theorems. *Phys. Rev.* **2018**, *8*. [CrossRef]
22. Kwon, H.; Kim, M.S. Fluctuation Theorems for a Quantum Channel. *Phys. Rev.* **2019**, *9*, 31029. [CrossRef]
23. Crooks, G.E. Nonequilibrium measurements of free energy differences for microscopically reversible Markovian systems. *J. Stat. Phys.* **1998**, *90*, 1481–1487. [CrossRef]
24. Lebowitz, J.L.; Spohn, H. A gallavotti-cohen-type symmetry in the large deviation functional for stochastic dynamics. *J. Stat. Phys.* **1999**, *95*, 333–365. [CrossRef]
25. Rao, R.; Esposito, M. Detailed fluctuation theorems: A unifying perspective. *Entropy* **2018**, *20*, 635. [CrossRef]
26. Esposito, M.; Lindenberg, K.; Van Den Broeck, C. Entropy production as correlation between system and reservoir. *New J. Phys.* **2010**, *12*. [CrossRef]
27. Jevtic, S.; Rudolph, T.; Jennings, D.; Hirono, Y.; Nakayama, S.; Murao, M. Exchange fluctuation theorem for correlated quantum systems. *Phys. Rev. Stat. Nonlinear Soft Matter Phys.* **2015**, *92*, 1–12. [CrossRef]

© 2020 by the authors. Licensee MDPI, Basel, Switzerland. This article is an open access article distributed under the terms and conditions of the Creative Commons Attribution (CC BY) license (http://creativecommons.org/licenses/by/4.0/).

Article
Probability Distributions with Singularities

Federico Corberi [1,2,*] **and Alessandro Sarracino** [3]

[1] Dipartimento di Fisica "E. R. Caianiello", Università di Salerno, via Giovanni Paolo II 132, 84084 Fisciano (SA), Italy
[2] INFN, Gruppo Collegato di Salerno, and CNISM, Unità di Salerno, Università di Salerno, via Giovanni Paolo II 132, 84084 Fisciano (SA), Italy
[3] Dipartimento di Ingegneria, Università della Campania "L. Vanvitelli", via Roma 29, 81031 Aversa (CE), Italy; alessandro.sarracino@unicampania.it
* Correspondence: corberi@sa.infn.it

Received: 28 February 2019; Accepted: 20 March 2019; Published: 21 March 2019

Abstract: In this paper we review some general properties of probability distributions which exhibit a singular behavior. After introducing the matter with several examples based on various models of statistical mechanics, we discuss, with the help of such paradigms, the underlying mathematical mechanism producing the singularity and other topics such as the condensation of fluctuations, the relationships with ordinary phase-transitions, the giant response associated to anomalous fluctuations, and the interplay with fluctuation relations.

Keywords: large deviations; phase transitions; condensation of fluctuations; fluctuation relations

1. Introduction

Quantitative predictions on the occurrence of rare events can be very useful particularly when these events can produce macroscopic effects on the system. This occurs, for instance, when a large fluctuation triggers the decay of a metastable state [1] leading the system to a completely different thermodynamic condition. Other examples with rare deviations producing important effects are found in many other contexts, as in information theory [2] and finance [3].

For a collective variable N, namely a quantity formed by the addition of many microscopic contributions, such as the energy of a perfect gas or the mass of an aggregate, typical fluctuations are regulated by the central limit theorem. Rare events, instead, may go beyond the theorem's validity and are described by large deviations theory [4,5] which, in principle, aims at describing the whole spectrum of possible fluctuations, no matter how large or rare they are, by means of their full probability distribution $P(N)$.

It has been found that, in many cases, $P(N)$ exhibits a singular behavior, in that it is non-differentiable around some value (or values) N_c of the fluctuating variable [3,6–39]. Such singularities have an origin akin to those observed in the thermodynamic potentials of systems at criticality. Indeed, a correspondence can be shown between $P(N)$ and the free energy of a companion system, related to the one under study by a duality map [4,34–36], which is interested by a phase-transition.

Recently, a great effort has been devoted to the characterization of these singular behaviors in the large deviation functions of different models where analytical results can be obtained. This has unveiled a rich phenomenology which shares common features. In most cases non-analyticities are a consequence of a particular condensation phenomenon denoted as condensation of fluctuations.

It occurs when a significant contribution to the fluctuations is built within a limited part of phase-space, or is provided by just one of the degrees of freedom of the system. This is analogous to what happens, for instance, in the usual condensation of a gas when it concentrates in a liquid drop, or in the well-known Bose–Einstein condensation, where the mode with vanishing wavevector contributes macroscopically. However, while usual condensation represents the typical behavior of the system, the condensation of fluctuations can only be observed when certain rare events take place.

Another interesting feature of systems with singular probability distributions can be their extreme sensibility to small perturbations. Usually, the properties of a system made of many constituents or degrees of freedom do not change much if some features of a single particle are slightly changed. This is true both for the average properties and for the fluctuations. For instance, neither the average energy of a gas nor its fluctuations change appreciably if the mass of one single molecule is increased a bit. This is simply because this particle is only one out of an Avogadro number. However, when condensation of fluctuations occurs, one can observe a giant response if the perturbed degree of freedom is exactly the one that contributes macroscopically to the fluctuation.

Singular probability distributions raise the question about the validity of the fluctuation relations (FRs). These relations have been extensively studied recently [40,41] because they reflect general symmetries of the deviations of certain quantities and are believed to contribute to a general understanding of non-equilibrium states. In particular, FRs connect the probability of observing events with a certain value N of the fluctuating variable, to the probability of the events associated to the opposite value $-N$. Among other open issues on the subject, one is represented by the case of singular fluctuations. Indeed, the singularity in N_c usually separates two regions where fluctuations have very different properties. For instance, on one side of N_c one can have a standard situation where all the degrees of freedom contribute, whereas on the other side fluctuations can condense and be determined by the contribution of a single degree. Clearly, if N is such that N and $-N$ fall on different branches of $P(N)$, namely on the two sides of N_c, the mechanism whereby an FR can be fulfilled must be highly non-trivial. In general, singular probability distributions may, or may not, exhibit the FR and a general understanding of this point is still not achieved.

This paper is a brief review devoted to the discussion of singular probability distributions where, without any presumption of neither completeness or mathematical rigor, we present examples of models where such non-analyticities show up, we highlight the mathematical mechanism producing condensation, and we discuss some relevant aspects related to the subject, such as those mentioned above. We do that in a physically oriented spirit, providing whenever possible an intuitive interpretation and a simple perspective. Non-differentiable probability distributions have been previously reviewed also in [42], where however the authors focus on different models and complementary aspects with respect to those addressed in this paper.

The paper is organized as follows. In Section 2 we recall some basic results of probability theory and introduce some notations. In Section 3 we present some models of statistical mechanics where non-differentiable probability distributions have been computed for different collective quantities. In Section 4 we illustrate in detail some phenomena related to the singular distribution function, mainly using the urn model as a paradigm, and discuss how similar behaviors arise in other systems. We also discuss the topic of the fluctuation relations. More specific features, such as giant response and observability, are then presented in Section 5, and, finally, some conclusions are drawn in Section 6.

2. Probability Distributions: Generalities

We consider a generic stochastic system, whose physical state is defined by the random variable x taking values on a suitable phase space. We will be mainly interested in the behavior of collective random variables, that are defined as the sum of a large number of microscopic random variables. For these quantities some general results can be derived [5]. As an example let us consider the sum $N = \sum_{j=1}^{M} x_j$ of a sequence of M random variables x_j, with empirical mean

$$\rho = \frac{N}{M} = \frac{1}{M} \sum_{j=1}^{M} x_j. \tag{1}$$

The quantities x_j can represent a sequence of states of a system (for instance, the position of a particle along a trajectory) or an ensemble of variables describing its microscopic constituents (e.g., the energies of the single particles of a gas). In the case of independent identically distributed variables, with expectation $\langle x \rangle$ and finite variance σ, one has that the empirical mean tends to $\langle x \rangle$ for large M, namely

$$\lim_{M \to \infty} p(\rho - \langle x \rangle < \epsilon) \to 1, \qquad (2)$$

where ϵ is a small quantity and hereafter $p(E)$ (also $P(E)$ or $\mathcal{P}(E)$) is the probability of an event E. The above equation represents the Law of Large Numbers.

As a further step, one can describe the statistical behavior of the small fluctuations of ρ around the average $\langle \rho \rangle$, $\delta\rho = \rho - \langle \rho \rangle$, introducing the quantity

$$z_M = \frac{1}{\sigma\sqrt{M}} \sum_{j=1}^{M} (x_j - \langle x \rangle), \qquad (3)$$

which, for very large M, and for $\delta\rho \lesssim O(\sigma/\sqrt{M})$, has the following distribution function

$$p(z_M = z) \simeq \frac{1}{\sqrt{2\pi}} e^{-\frac{z^2}{2}}. \qquad (4)$$

This result is the central limit theorem (CLT), that holds also in the case of weakly correlated variables.

More in general, fluctuations of arbitrary size of the quantity ρ can, under certain conditions, be characterized by the large deviation principle (LDP)

$$p(\rho = y) \sim e^{-MI(y)}, \qquad (5)$$

where $I(y)$ is the so called rate function. When $p(\rho)$ has a single absolute maximum (in $\langle \rho \rangle$), the rate function is positive everywhere but for $y = \langle \rho \rangle$, where it vanishes. It is easy to obtain the CLT Equation (4) from the LDP Equation (5) by expanding up to second order the function $I(y)$ around $\langle \rho \rangle$. However, as we will discuss in detail below, there are interesting cases where the LDP in the form Equation (5) is not satisfied.

A simple example where LDP holds and the rate function can be easily computed is obtained by considering $\{x_j\}$ as dichotomous variables taking the value $+1$ with probability q and -1 with probability $1-q$. Then, using the Stirling approximation, one obtains the explicit expression for the rate function:

$$I(y) = \frac{1+y}{2} \ln \frac{1+y}{2q} + \frac{1-y}{2} \ln \frac{1-y}{2(1-q)}. \qquad (6)$$

Expanding Equation (6) around the mean $\langle y \rangle = 2q - 1$ one has the CLT

$$I(y) \simeq \frac{(y - \langle y \rangle)^2}{2(1 - \langle y \rangle)}. \qquad (7)$$

3. Singular Probability Distributions: Examples

As far as small deviations of a collective variable are considered, the associated probability distribution is usually regular, being a Gaussian when the hypotheses of the CLT are satisfied. Moving to the realm of large deviations, instead, can hold surprises as, for instance, the emergence of non-analyticities. Before deepening the meaning and the bearings of the singular behavior, in this section we first itemize some examples of systems where it has been observed. We will then study it in more detail in some specific models in the following sections.

3.1. Gaussian Model

The Gaussian model is a reference model of statistical mechanics. An order-parameter field $\phi(\vec{x})$ (which in the magnetic language can be thought of as a local magnetization at site $\vec{x}$) is ruled by the following Hamiltonian

$$\mathcal{H}[\varphi] = \frac{1}{2} \int_V d\vec{x} \, [(\nabla \varphi)^2 + r \varphi^2(\vec{x})], \tag{8}$$

where $r > 0$ is a parameter and V the volume. This simple model can be exactly solved and has a rather trivial phase-diagram without phase transitions.

Let us consider the collective variable

$$N[\varphi] = \int_V d\vec{x} \, \varphi^2(\vec{x}), \tag{9}$$

namely the order parameter variance, and its density $\rho = N/V$. Its probability distribution was computed analytically in [34–36]. The (negative) rate function of this quantity, evaluated in equilibrium at a given temperature T, is plotted in Figure 1. The curve has a maximum in correspondence to the most probable value, where $I(\rho)$ vanishes. Far from such maximum, in the large deviations regime, the rate function exhibits a singularity (marked with a green dot) at $\rho = \rho_c$. In this point the third derivative of the rate function has a discontinuity [34–36]. The existence of such a singularity is related to the fact that, as we will discuss later, fluctuations with $\rho > \rho_c$ have a different character with the respect to the ones in the region $\rho < \rho_c$ where the average, or typical, behavior of the system (i.e., the most probable value of ρ) is located.

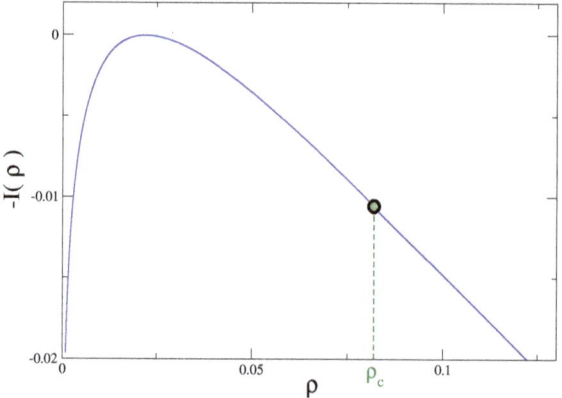

Figure 1. The (negative) rate function $I(\rho)$ of the variance N of the order parameter field in the Gaussian model in $d = 3$, with $r = 1$, in equilibrium at the temperature $T = 0.2$.

3.2. Large-$\mathcal{N}$ Model

Another reference model of statistical mechanics is the description of a magnetic system in terms of the Ginzburg–Landau Hamiltonian

$$\mathcal{H}[\varphi] = \frac{1}{2} \int_V d\vec{x} \, \left[(\nabla \varphi)^2 + r \varphi^2(\vec{x}) + \frac{g}{2\mathcal{N}} (\varphi^2)^2 \right], \tag{10}$$

where the $\mathcal{N}$-components vectorial field φ has a meaning similar to that of the Gaussian model, and $r < 0$ and $g > 0$ are parameters. In the large-$\mathcal{N}$ limit (sometimes also denoted as the spherical limit) the model is exactly soluble. There is a phase transition at a finite critical temperature T_c separating a paramagnetic phase for $T > T_c$ from a ferromagnetic one at $T < T_c$.

The probability distribution of the energy $N(t, t_w) = \mathcal{H}[\varphi, t] - \mathcal{H}[\varphi, t_w]$ exchanged by the system in a time interval $[t_w, t]$ with a thermal bath was computed exactly in [37]. The (negative) rate function of the intensive quantity $\rho(t, t_w) = N(t, t_w)/V$ is shown in Figure 2. This figure refers to the case of a system quenched from a very high temperature to another $T < T_c$. Also in this case there is a singularity corresponding to a certain value the quantity $\rho(t, t_w) = \rho_c$ where the third derivative has a discontinuity, and this reflects a different mechanism of heat exchanges for $\rho < \rho_c$ and for $\rho > \rho_c$.

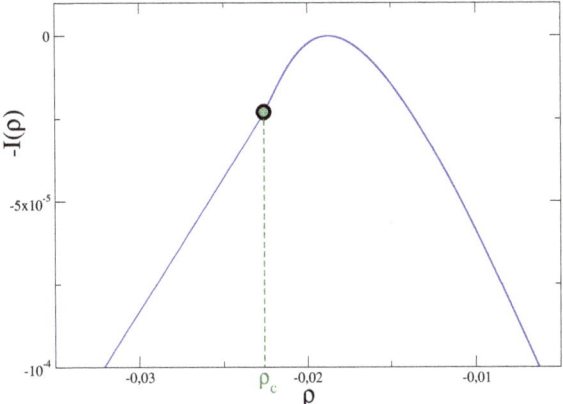

Figure 2. The (negative) rate function $I(\rho)$ of the probability distribution $P(N)$ of the energy N exchanged by the large-$\mathcal{N}$ model in $d = 3$, with $g = -r = 1$, with the environment after a quench to zero temperature.

3.3. Urn Model

Let us consider a set of integer variables $n_i \geq 0$ ($i = 1, \ldots, M$) equally distributed in such a way that the probability of having a certain value n of n_i is

$$p(n) = \zeta^{-1}(n+1)^{-k}, \tag{11}$$

where ζ is a normalization constant and k a parameter. One can think of having M urns, each of them hosts a quantity n_m of particles taken with probability Equation (11) from a reservoir. This setting is appropriate to describe a wealth of situations in many areas of science, from network dynamics to financial data. The probability distribution of the total number of particles

$$N = \sum_{m=1}^{M} n_m \tag{12}$$

was studied for large M in different contexts [14,17,21–23,43]. The (negative) rate function is shown in Figure 3. Also in this model it is found that, if $k > 2$, there is a singularity at $\rho = \rho_c$, that in this particular case coincides with the average value $\langle \rho \rangle$. Notice that in this case, at variance with the previous examples, the rate function vanishes in the whole region $\rho \geq \rho_c$. This is due to the fact that $P(\rho)$ has a weaker dependence on M with respect to the exponential one of Equation (5), and hence the LDP is violated for $\rho > \rho_c$. We will comment later on that.

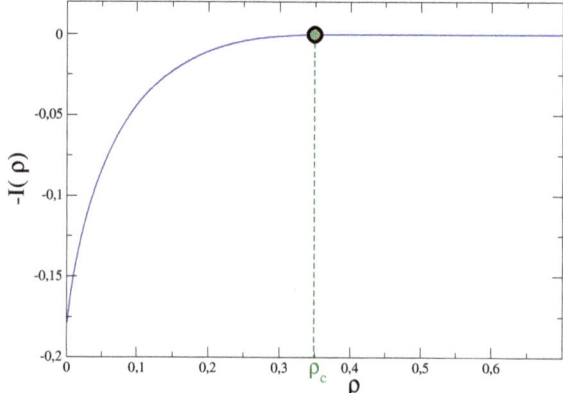

Figure 3. The rate function $I(\rho)$ of the probability distribution $P(N)$ of the total number of particles N in the urn model with $k = 3$.

3.4. Stochastic Maxwell–Lorentz Particle Model

The so-called stochastic Maxwell–Lorentz gas [44,45] consists of a probe particle of mass m whose velocity v changes due to the collisions with bath particles, of mass M at temperature T, and due to the acceleration produced by an external force field $\mathcal{E}$. Collisions with the scatterers change instantaneously the probe's velocity from v to v' and we assume the simple collision rule $v' = V$, where V is the velocity of the scatterer, drawn from a Gaussian distribution:

$$P_{\text{scatt}}(V) = \sqrt{\frac{M}{2\pi T}} e^{-\frac{MV^2}{2T}}. \tag{13}$$

The scatterers play the role of a thermal bath in contact with the probe particle. This model is a particular case of a more general class of systems studied in [44,45]. During a time τ between two consecutive collisions, the probe performs a deterministic motion under the action of the field $\mathcal{E}$. We assume that the duration of flight times τ is exponentially distributed $P_\tau(\tau) = \frac{1}{\tau_c} \exp(-\tau/\tau_c)$ and independent of the relative velocity of the particles. The system reaches a non-equilibrium stationary state characterized by a total entropy production Δs_{tot}, associated with the velocity $v(t)$, defined as

$$\Delta s_{\text{tot}}(t) = \ln \frac{P(\{v(s)\}_0^t)}{P(\overline{\{v(s)\}_0^t})}, \tag{14}$$

where $P(\{v(s)\}_0^t)$ and $P(\overline{\{v(s)\}_0^t})$ are, respectively, the pdf of a path $\{v(s)\}_0^t$ spanning the time interval $[0, t]$ and of the time-reversed path $\overline{\{v(s)\}_0^t} = \{-v(t-s)\}_0^t$ [46]. This fluctuating quantity takes contributions at any time and is therefore extensive in t. In this example it plays the role of the collective variable N, and t plays the role of the number M of elements contributing to it.

The rate function $I(\rho)$ of the quantity $\rho = \Delta s_{\text{tot}}/t$ was studied in [11] by means of numerical simulations for finite times and analytically in the limit $t \to \infty$. This quantity is shown in Figure 4, where $\rho_c = m\tau_c \mathcal{E}^2/\theta$, with $\theta = Tm/M$ playing the role of an effective temperature [47]. Also in this case, as for the urn model, $I(\rho)$ vanishes and $P(\Delta s_{\text{tot}})$ does not satisfy a standard LDP for $\rho > \rho_c$. Indeed it can be shown that the far positive tail of $P(\Delta s_{\text{tot}})$ scales exponentially with $\sqrt{t}$ rather than with t [11], how it should be if the LDP (Equation (5)) holds. Recently, the nature of the singularities in $I(\rho)$ and their physical meaning have been thoroughly discussed in a similar model in [48], where the observed non-analytical behaviors have been related to a first-order dynamical phase transition.

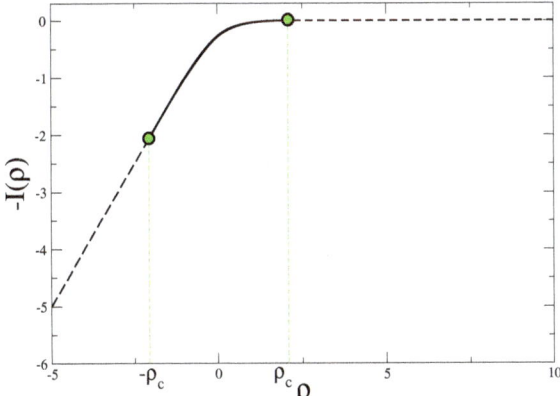

Figure 4. The rate function $I(\rho)$ of the quantity $\rho = \Delta s_{tot}/t$ for the Maxwell–Lorentz gas model [11], computed analytically in the limit $t \to \infty$.

3.5. Some Other Models

We have discussed above some models where a singular probability distribution was found. All these cases can be grouped into two classes: the first contains the cases where the rate function is well defined, although it contains some non-analyticity point. The examples of Sections 3.1 and 3.2 behave in this way. The second class is the one represented by the urn model, where the probability distribution is still singular, but the the rate function is not defined in a certain region (that is to say it vanishes identically). The Maxwell–Lorentz gas is an example where the two behaviors are exhibited in different regions of the fluctuation spectrum.

Beyond the cases discussed before, other examples of singular behavior include the probability distribution of the work done by active particles [38], of the heat exchanged by harmonic oscillators during a quench with a thermal bath [39], of the magnetization in the spherical model [6,7], of the displacement of a Brownian walker with memory [10], of the work done in a quantum quench [12], and many others [4,25–33].

We also mention the case where the singularity appears as a "kink" at zero in the probability distribution, showing a linear regime for negative values. This behavior has been observed in the distribution of the entropy production and of other currents for a driven particle in periodic potentials [49–51], in a molecular motor model, described in [52], and in the experimental results reported in [53], where the large deviation function of the velocity of a granular rod was measured. In general, the presence of the kink can be related to different physical mechanisms [54], such as intermittency [55], detailed fluctuation theorem [56], and dynamical phase transitions [57]

4. General Features of Singular Probability Distributions

In this section we will discuss some general properties of singular probability distributions observed in the different models mentioned above, focusing on the common physical interpretation and on the underlying mathematical structure.

4.1. Duality

The singular behavior of the probability distribution seen in the examples of the previous section has an interpretation akin to the occurrence of phase transitions in ordinary critical phenomena. In order to discuss this point we can refer to the Gaussian model as a paradigm. The partition function is

$$Z = \int \delta\varphi \, \mathcal{P}(\varphi), \tag{15}$$

where $\mathcal{P}$ is the probability of microscopic configurations as specified by the field φ. For instance, in a canonical setting it is $\mathcal{P}(\varphi) = Z^{-1}\exp[-\beta\mathcal{H}(\varphi)]$, where β is the inverse temperature $\beta = 1/(k_B T)$; in this case Z depends on T and V, the volume. On the other hand the probability of the collective variable N of Equation (9) can be written as

$$P(N) = \int \delta\varphi\, \mathcal{P}(\varphi)\, \delta\left(\int_V d\vec{x}\,\varphi^2(\vec{x}) - N\right). \tag{16}$$

In view of Equation (15), one can recognize Equation (16) as a partition function as well. However this is not the partition function of the original model that is, in this example, the Gaussian one. Instead, $P(N)$ in Equation (16) can be interpreted as the partition function of a dual system that can be obtained from the original one upon removing all the configurations such that the argument of the delta function in Equation (16) does not vanish. In other words, this is the model one arrives at upon constraining configurations in a certain way. In this case the requirement is that the variance of φ must equal a given value N. Such a system, a Gaussian model with a constraint on the variance, is the spherical model of Berlin and Kac [58].

The equilibrium properties of the spherical model are exactly known. For fixed N, there is a phase-transition at a critical temperature T_c, from a disordered phase for $T \geq T_c$ to an ordered one below T_c. Equivalently, still in the Berlin–Kac model, if one keeps T fixed, the transition occurs changing the variance $N[\varphi]$ defined in Equation (9) upon crossing a critical value N_c. The ordered phase is found for $N > N_c$, in this case. The presence of such a phase transition crossing N_c determines a singularity of the partition function $P(N)$ of the spherical model (Equation (16)) at $N = N_c$. However the same quantity $P(N)$ is also the probability distribution of the quantity $N[\varphi]$ in the context originally considered, the Gaussian model. This explains what one observes in Figure 1. N_c is the value of N marked by a dot in this figure, where the singular behavior shows up.

This dual interpretation of $P(N)$, as a probability distribution of a collective variable in the original model, or as a partition function in a dual model, may help to understand why singularities are manifested in the probability distributions. Indeed, if one asks the question: why a simple model without phase transitions, such as the Gaussian model, exhibits a non trivial singularity in the probability distribution $P(N)$, the answer can be that, although the original model is quite simple, the dual one is far from being trivial, with a phase-transition induced by the presence of the constraint. This generates anomalous behavior in the fluctuation spectrum of the original model.

We have discussed the fact that imposing a constraint to the Gaussian model we change the system into a dual one that is interested by a phase transition, since this is the spherical model. Is this an isolated example or has this feature some generality? The answer is that it occurs quite often. Besides the above mentioned spherical model, another well known example where the same mechanism is at work is the perfect boson gas. There is no phase transition in a gas with a non conserved number N of bosons, as in the case of photons, but if the number N of particles is fixed Bose–Einstein condensation happens. The partition function of the conserved bosons, for a given volume and temperature, has a singularity at a certain value of the boson number $N = N_c$ (or density). This singularity corresponds to the critical number of particles below which the condensed phase develops. According to the duality principle discussed above, this implies that the probability distribution of the number of bosons in a system of, say, photons, where this number is allowed to fluctuate, will be singular at the same value N_c of the random variable N [33]. The very urn model is another instructive example. One can consider a model, dual to the one discussed in Section 3.3, where the total number of balls is conserved [21]. Marbles can only be exchanged among boxes and their density ρ is an external control parameter. This model is known to be interested, for $k > 2$, by a phase transition crossing $\rho = \rho_c$. Notice that, since ρ is a control parameter, having $\rho > \rho_c$ in this dual model is not a rare event (as in the model introduced in Section 3.3). A similar situation is found in related models such as the zero range process [18,21,28].

4.2. Condensation

In order to see how singularities may come about in another perspective we will discuss the phenomenon in the framework of the urn models, where the physical meaning is probably more transparent in term of a condensation mechanism. Something similar occurs also in the other models considered in Section 3, regardless of the fact that the rate function is well defined or not.

Let us consider the conditional probability $\pi(n, N, M)$ that one of the M a priori equivalent urns contains n particles, given that there are N particles in the whole system. This quantity can be evaluated exactly and is shown in Figure 5 (normalized by its value in $n = 0$ to better compare curves with different N in a single figure). Let us discuss its properties. First of all π vanishes for $n > N$, since it is impossible that an urn contains more particles than the whole system. Secondly, for small n one has $\pi(n, N, M) \propto p(n)$ (dotted green line in Figure 5). This means that, as far as very few particles are stored in the tagged urn, the condition on the total number of balls is irrelevant.

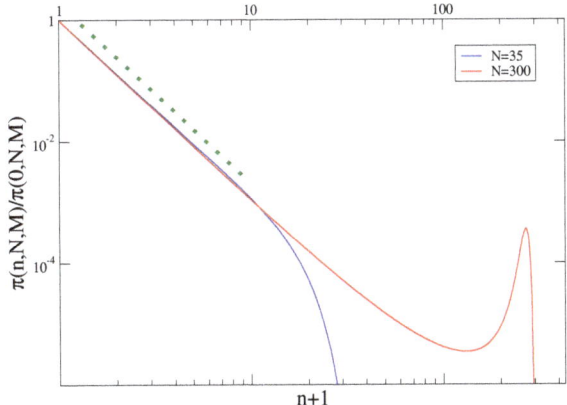

Figure 5. The function $\pi(n, N, M)$ is plotted, for $k = 3$ and $M = 100$, against $n + 1$ for two values of N: $N = 35$, corresponding to a case without condensations, and $N = 300$, corresponding to a condensed situation. The dotted green curve is the power-law x^{-k}.

More interestingly, at large n, $n \lesssim N$, different behaviors are observed in the region of relatively small N, and in the one with relatively large N, exemplified by $N = 35$ and $N = 300$, respectively, in Figure 5. In the former case π is exponentially damped at large n, meaning that accommodating many particles in a single urn is probabilistically very unfavorable. In the latter case there is a peak at a value of n or of order N. This means that a significant fraction of the total number of particles is located in a single urn. This is the condensation phenomenon. (We will see in the next section that in this particular model occurs when $k > 2$ and for sufficiently large densities). The essence of a condensation phenomenon is that a given quantity is not fairly distributed among many degrees of freedom, but is concentrated in a single one. This is particularly clear in the urn model, where one particular urn contains a macroscopic fraction of balls.

One easily realizes that something similar occurs in the other example models discussed in Section 3. For instance, in the model of Section 3.1, writing $N[\varphi]$ in terms of the Fourier components $\varphi_{\vec{k}}$ of the field φ as

$$N[\varphi] = \frac{1}{V} \sum_{\vec{k}} |\varphi_{\vec{k}}|^2, \qquad (17)$$

one can show that while for $N \leq N_c$ (or equivalently $\rho \leq \rho_c$) all the Fourier components add up to realize the sum in Equation (17) in a comparable way, for $N > N_c$ the term with $k = 0$ alone provides the most important contribution to the sum. A similar mechanism, with the dominance of the $k = 0$ term, is also at work in the example of Section 3.2. In the Maxwell–Lorentz particle model

(Section 3.4) one has that normal entropy fluctuations are formed by the addition of many contributions associated to many short flights of the probe particle. However above the critical threshold ρ_c they are associated to a single event which is responsible for a macroscopic contribution to the entropy production. This event is a long flight of the probe particle with no collisions with the scatterers. For more details and a very accurate analytical description of these kinds of behaviors in a similar system re-framed in the context of active particles, see the recent work [48].

4.3. Mathematical Mechanism

In the previous section we have discussed the phenomenon of condensation on physical grounds. In this section we show the underlying mathematical mechanism. We will give a description as simple as possible, without presumption of mathematical rigor, in the framework of the urn model.

The probability distribution of the total number of particles N reads

$$P(N,M) = \sum_{n_1,n_2,\ldots,n_M} p(n_1)p(n_2)\ldots p(n_M) \delta_{\sum_{m=1}^M n_m, N}, \tag{18}$$

where $\delta_{a,b}$ is the Kronecker function and in the leftmost sum the variables $n_1, n_2, \ldots, n_M$ run from 0 to ∞. Using the representation

$$\delta_{a,b} = \frac{1}{2\pi i} \oint dz\, z^{-(b-a+1)} \tag{19}$$

of the δ function one arrives at

$$P(N,M) = \frac{1}{2\pi i} \oint dz\, e^{M[\ln Q(z) - \rho \ln z]}, \tag{20}$$

where

$$Q(z) = \sum_{n=0}^{\infty} p(n) z^n, \tag{21}$$

and we have confused $\frac{N+1}{M}$ with $\rho = N/M$ for large M. Still for large M, the integral in Equation (20) can be evaluated by the steepest descent method as

$$P(N,M) \simeq e^{-MR(\rho)}, \tag{22}$$

where

$$R(\rho) = -\ln Q[z^*(\rho)] + \rho \ln z^*(\rho), \tag{23}$$

with z^* the value of z for which the argument in the exponential of Equation (20) has its maximum value. This in turn is given by the following implicit saddle-point equation

$$\Theta(z^*) = \rho, \tag{24}$$

where

$$\Theta(z^*) = z^* \frac{Q'(z^*)}{Q(z^*)}. \tag{25}$$

Let us study this equation. Clearly, it must be $z \leq 1$ in order for the sums hidden in Q and Q' to converge. It can also be easily seen that $\Theta(0) = 0$ and that this function increases with z up to

$$\Theta(1) = \begin{cases} \infty, & k \leq 2 \\ \Theta_M, & k > 2, \end{cases} \tag{26}$$

where Θ_M is a finite positive number. The function $\Theta(z)$ is shown in Figure 6, for two values of the parameter k. As it is clear from this figure, for $k > 2$ the saddle point Equation (24) admits a solution only for $0 \leq \rho \leq \rho_c = \Theta(1)$. It is trivial to show that $\rho_c \equiv \langle \rho \rangle = \sum_n np(n)$. However nothing prevents fluctuations with $\rho > \langle \rho \rangle$ to occur. How can we recover the model solution for $\rho > \langle \rho \rangle$? We know that for such high densities urns are no longer equivalent: there is one—say the first—which hosts an extensive number of particles and condensation occurs. In a physically oriented approach, we can take into account this fact by writing, in place of Equation (18), the following

$$P(N,M) = M \sum_{n_1=0}^{\infty} p(n_1) \sum_{n_2,n_3,\ldots,n_M} p(n_2) p(n_3) \ldots p(n_M) \delta_{\sum_{m=2}^{M} n_m, N-n_1}. \tag{27}$$

The factor M in front of the r.h.s. stems from the fact that there are M ways to chose the urn (denoted as 1) to be singled out. Repeating the mathematical manipulations as in Equations (18) and (20), but only on the sum $\sum_{n_2,n_3,\ldots,n_M} \ldots$, one arrives at

$$P(N,M) = \frac{M}{2\pi i} \sum_{n_1=0}^{\infty} p(n_1) \oint dz\, e^{M[\ln Q(z) - (\rho - \frac{n_1}{M})\ln z]}. \tag{28}$$

Evaluating the integral with the steepest descent method, the saddle point equation is now

$$\Theta(z^*) = \rho - \frac{n_1}{M}. \tag{29}$$

Notice that in a normal situation, where condensation does not occur, in the thermodynamic limit where $M \to \infty$ with fixed ρ, the typical number of particles in a single urn does not depend on the number of urns. Therefore the last term on the r.h.s. of Equation (29) is negligible and one goes back to the previous saddle point Equation (24). However, when condensation occurs (i.e., with $k > 2$ and $\rho > \langle \rho \rangle$) the only possibility to close the model equations is to have the last term in Equation (29) finite. In conclusion one has

$$\begin{cases} z^* < 1, & \frac{n_1}{M} \simeq 0 \quad \text{no condensation} \\ z^* = 1, & \frac{n_1}{M} = \rho - \langle \rho \rangle \quad \text{condensation.} \end{cases} \tag{30}$$

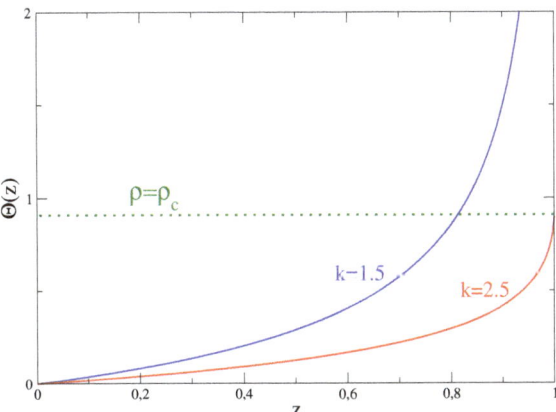

Figure 6. The function $\Theta(z)$ is shown for $k = 1.5$ and $k = 2.5$.

Clearly we are in the presence of a phase-transition resembling the ferro-paramagnetic or the gas–liquid transitions. There are two phases with qualitatively different behaviors. However, at variance with usual phase transitions, here the parameter producing the transition is not an external one that can be varied at will, but the value of the spontaneously fluctuating variable N. Another difference with usual phase transitions is the fact that here there is no interaction among urns. Despite that, urns are not completely independent due to the constraint over the number of particles represented by the Kronecker function in Equations (18) and (27). This constraint can be regarded as an effective interaction determining the transition (it can be easily seen, in fact, that without such conservation there is no transition).

Notice that it is $n_1/M = 0$ in the normal phase and $n_1/M \neq 0$ in the condensed phase, therefore this quantity represents the order parameter of the transition. Despite the fact that a priori the system (i.e., the Hamiltonian) is invariant under a permutation of the urns, namely all boxes are equal, this property is not shared by the physical realization of the actual state of the system when condensation occurs, since one urn behaves very differently from the others. We are in the presence of spontaneous symmetry breaking.

As a final remark, let us note that the phenomenon of condensation in the sum of many identically distributed variables is not specific to an algebraic decay of $p(n)$, or to the discrete value of the variable n. Indeed it is found [31] that it occurs provided that $\sum_n np(n) < \infty$. Condensation in the presence of a stretched exponential $p(n)$, for instance, has been discussed in [59,60]. Finally, we mention the fact that in the context of Lévy flights the phenomenon of condensation is usually referred to as the big jump principle [61].

4.4. Fluctuation Relation

The Fluctuation Relation is one of the few general results of non-equilibrium statistical mechanics, expressing an asymmetry property of the fluctuations of some extensive (in time or in number of degrees of freedom) quantities N [40]. The FR reads

$$\frac{P(N/M = \rho)}{P(N/M = -\rho)} = e^{cM\rho + o(M)}, \qquad (31)$$

where c is a constant, and $o(M)$ stands for sub-linear corrections in M. Usually, the exponential form of the FR is related to two properties of $P(N/M)$: (i) it satisfies a LDP Equation (5), and (ii) the rate function $I(\rho)$ has the symmetry:

$$I(-\rho) - I(\rho) = c\rho. \qquad (32)$$

These two conditions, with $I(\rho)$ different from 0 and ∞, are known to be sufficient for N/M to satisfy a FR (see, e.g., [4] and references therein).

It is interesting to consider the validity of an FR in the case of probability distributions with singularities. First, let us note that, when the singularity appears in zero, as in the case of the "kink" mentioned in Section 3.5, then the validity of an FR is clearly not affected by the singularity. More in general, the FR can also be satisfied by a pdf for which a standard (namely, with a leading exponential scaling in M) LDP does not hold. This can be observed for instance in the driven Maxwell–Lorentz gas described in Section 3.4. In this model it has been shown [11] that the entropy production calculated over a time t satisfies an FR, even though the far positive tail of its pdf scales exponentially with $\sqrt{t}$ rather than t. In this case the validity of the FR can be exploited to extract some information on the behavior of the probability distribution in the regions where the stretched-exponential scaling takes place.

The FR Equation (32) in the presence of a singular rate function has been also observed [37], besides the already mentioned Maxwell-Lorentz case , in some large time limit for the exchanged heat, in the large-$\mathcal{N}$ model of Section 3.2. More recently, it has been shown [39] that the rate function of the heat exchanged by a set of uncoupled Brownian oscillators with the thermostat during a non-stationary relaxation process does not satisfy an FR in the form Equation (31). Although, even in this case, the rate function shows a singular behavior in the limit of a large number of degrees of freedom, the lack of a standard FR is not necessarily related to the presence of the singularity.

5. Some Peculiarities of Singular Distributions

5.1. Giant Response

Generally, the behavior of a collective quantity such as the empirical mean Equation (1) is not substantially altered if, for large M, the properties of only one out of M variables is slightly modified. For instance, one does not expect to observe any significant change in the thermodynamic properties of a gas of identical molecules if one is replaced with another of a different substance. This is because the collective properties are determined by the synergic contribution of a huge number M of constituents, and hence the features of a single molecule are negligible. This is true not only for the typical properties but also for the fluctuation distribution. However, the situation can be dramatically different when singular probability distributions enter the game.

Let us show this with the prototypical example of the urn model. We consider a slightly modified version of the model defined in Section 3.3, where a single variable, say n_ℓ, is distributed as in Equation (11) but with an exponent k_ℓ that may be different from the one, k, of all the remaining ones. Let us now look at Equations (28)–(30). In a situation where condensation does not occur, as we remarked earlier, the effect of a single variable is negligible, the first line of Equation (30) applies and hence $\frac{n_m}{M} \simeq 0$, for any m. On the other hand, in the presence of condensation, the second line of Equation (30) holds. In the case of equally distributed variables condensation occurs with equal probability in any of the urns. However, if the ℓ-th variable behaves differently, one has to understand if the condensing variable could be the ℓ-th, or any of the remaining ones. Both the cases can occur, depending on the values of the exponents k and k_ℓ.

For $k_\ell > k > 2$ (the latter inequivalence being needed for condensation) it is $p(n_\ell = n) \ll p(n_m = n)$ for large n (with $\ell \neq m$). Hence the condensation phenomenon, which occurs by letting a huge amount of particles occupy a single urn, is unfavoured in the ℓ-th urn. The situation in this case is analogous to the one discussed before with equally distributed variables, i.e., with $k_\ell = k$. However for $k > k_\ell > 2$ the opposite occurs, the condensing variable is the ℓ-th. Hence Equation (28) applies with n_1 replaced by n_ℓ. One sees from Equation (28) that, when condensation occurs, $P(N, M)$ is proportional to $p(n_\ell)$. Since $k_\ell \neq k$, $P(N, M)$ turns out to be different from the one found for equally distributed variables. Hence, in this case, an even small change of the properties of a single variable can trigger the form of the probability distribution of the collective variable N, a fact that is sometimes referred to as giant response.

This is illustrated in Figure 7. Here $P(N, M)$ is compared for three different choices of the exponents k, k_ℓ. The continuous blue curve with asterisks refers to the case (i) with identically distributed variables with $k_\ell = k = 3$. Similarly, the dot-dashed green curve with squares corresponds to the situation with (ii) $k_\ell = k = 6$. Instead, the dashed-magenta curve with circles corresponds to non-identically distributed variables with (iii) $k_\ell = 3$ and $k = 6$. Notice that in the region to the left of the maximum, where condensation does not occur (because in this region $\rho < \langle \rho \rangle$), the curves of the cases (ii) and (iii) coincide. This nicely shows that in the absence of condensation the shift of the properties of a single variable does not influence the collective behavior of the system. For $\rho > \langle \rho \rangle$, on the other hand, the form of P drastically changes in going from (ii) to (iii), namely by perturbing the properties of one single variable. Even more impressive, the slope of the curve for case (iii) is the same as that of case (i), showing that this feature is dictated by the sole properties of the variable, n_ℓ, which in case (iii) behaves as in (i).

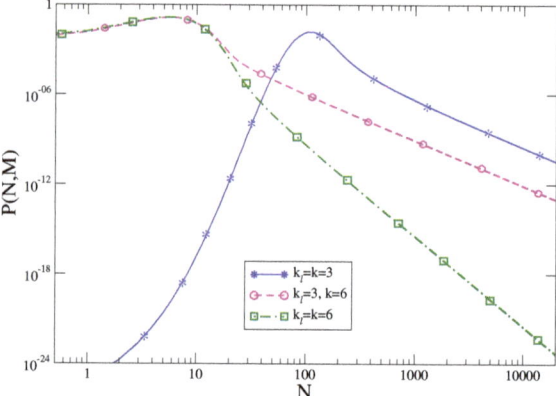

Figure 7. P is plotted for $M = 333$ and the three different choices (see text) (i) $k_\ell \equiv k = 3$, continuous blue with asterisks, (ii) $k_\ell \equiv k = 6$, dot-dashed green with squares and (iii) $k_\ell = 3$, $k = 6$, dashed magenta with a circles.

5.2. Development of a Singular Fluctuation

We have seen in Section 4.1 that a singularity in the probability distribution can be interpreted as a phase transition occurring at a critical value of ρ, playing the role of a control parameter. The analogy can be pushed a step further. When a system is prepared in a certain equilibrium state and then a control parameter is changed as to make it cross a phase transition, the ensuing dynamics can be slow and characterized by a dynamical scaling symmetry associated most of the times with an ever growing length scale [62–65]. Typical examples are magnets and binary systems quenched across the critical temperature, and glassy systems.

Building on the analogy above, one might expect something similar to happen if one prepares a system with a singular $P(N)$ in a state such that the fluctuating collective variable N takes a definite value N_0 on one side (say the left) of the critical value N_c where the singularity takes place. If the system is then left to evolve freely, all possible fluctuations will take place, including those on the other side (say the right) of the singularity. Due to the duality principle, this process should occur in a way akin to the kinetics of a system brought across a phase-transition. Hence slow evolution and dynamical scaling should be observed. This has actually been shown to be the case, as we discuss below.

Upon supplementing the urn model of Section 3.3 with a kinetic rule allowing the system to exchange single particles with an external reservoir in such a way that the stationary occupation probability of any urn is given by Equation (11), one can solve exactly [43] the evolution of a system whose initial state is such that condensation is not present. In the following we will discuss the case in which the initial value of the density is $\rho = \langle \rho \rangle$. Starting from this configuration, corresponding to an initial form $P(N, M, t = 0)$ of the probability distribution of the collective variable, the system will evolve as to produce all the allowed fluctuations. Hence $P(N, M, t)$ becomes time-dependent. Clearly, for long times it is expected to approach the stationary value $P(N)$, with the singular behavior already discussed. This curve is plotted in Figure 8, with a dotted green line.

In this figure one sees that the time evolution of the probability $P(N, M, t)$ towards this asymptotic form is much different on the two sides of the singularity. For $N < \langle N \rangle$, in the normal region without condensation, the evolution is fast and the asymptotic form of the probability is attained at relatively short times. Indeed, already the red curve for $t = 1.2 \times 10^6$ is indistinguishable from the stationary form and increasing time does not change anything. Conversely, the evolution is slow in the condensing region for $N > N_c$. Here one sees that, at any time, the asymptotic form is only attained up to a value $N = v(t)$, beyond which $P(N, M, T)$ drops much faster than what expected asymptotically. It can be shown that $v(t)$ grows indefinitely in an algebraic way, much in the same way as a characteristic growing length does in systems quenched across a phase transition. In addition, a dynamical scaling symmetry can be shown to be at work also in this case. The origin of this slow kinetics is clearly due to the difficulty to condense a huge amount of particles in a single urn by exchanging single particles with the reservoir.

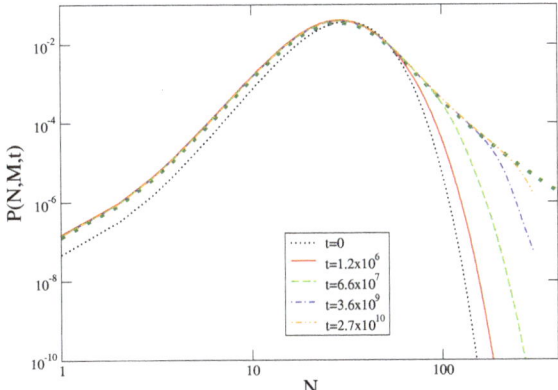

Figure 8. The probability $P(N, M, t)$ with $k = 3$ is plotted against N with double logarithmic scales for different times (see key), exponentially spaced. The dotted green line is the asymptotic form.

5.3. Observability

In the previous sections we discussed some peculiar properties of singular distribution functions. A natural question is if such features can be observed in practical situations. Indeed, the non-analycities of the probability distributions are observed in the regime of large deviations, namely outside the range of small fluctuations which are generally described by the central limit theorem and are more likely to be observed.

To make more clear this point let us make reference to the Gaussian model and, specifically, to Figure 1. In this case, in order to detect singular deviations, $\rho = \rho_c$ must be exceeded. Namely, the system has to move quite far from the most likely observed value—the maximum of the distribution. If the LDP Equation (5) holds (it does so in this model) the possibility to observe such a large fluctuation is extremely small already for moderately large values of the number of constituents M (or volume V), due to the exponential damping in M expressed by Equation (5). But the situation is different if the LDP is violated. This occurs, for instance, in the urn model or in the Maxwell–Lorentz gas, in the fluctuation range where the rate function vanishes. In the former model one can easily check from Equations (28)–(30) that the LDP is obeyed in the non-condensing regime but it is violated when condensation occurs. In fact, it is trivial to see that with $z^* - 1$ the saddle point evaluation of the integral in Equation (28) gives an exponential with an argument that is identically vanishing. As a consequence fluctuations away from the average are no longer damped exponentially in M, but only as M^{1-k} (keeping ρ fixed). This is why the rate function of the model vanishes in the whole sector $\rho > \rho_c$ where condensation occurs (see Figure 3), despite the fact that $P(N, M)$ decays for $\rho > \langle \rho \rangle$, as it can be seen in Figure 7. Due to this much softer decay, there is a better chance to observe singular fluctuations in this model than in others, e.g., the Gaussian model, where the LDP holds. A similar situation, with LDP violations, is observed also in the Maxwell–Lorentz particle model (for $\rho = \Delta s_{tot}/t > \rho_c$) [11] and in Bose–Einstein condensates [33].

6. Summary and Conclusions

In this paper we have shortly reviewed the issue of probability distributions characterized by non-analyticities. Naively, this feature could be considered as a rare manifestation of curious mathematical pathologies occurring in scholarly model with uncertain relations to the physical world.

In reality, singular probability distributions have been shown analytically to occur in very simple and fundamental models of statistical mechanics, such as the Gaussian one, and not only in weird non-equilibrium states but also in equilibrium. Furthermore, they have been detected in numerical simulations and, most importantly, also in real experiments. This widespread occurrence points towards an underlying general mechanism for the development of singularities in the fluctuation probability. This paper has been conceived in order to highlight and discuss, at a simple and physically oriented level, at least some of such general features.

In the first part of the paper, after recalling basic and general concepts of probability theory we have reviewed some models where singular fluctuation spectra have been observed. These range from the aforementioned Gaussian model to the spherical limit of a ferromagnet, from the so-called urn model to a description of the Maxwell–Lorentz gas. In all these cases the deviations of certain collective observables are described by non-analytical probability distributions, which, in the case when LDP holds, are characterized by the presence of exponential branches.

The non-analytical behavior has been interpreted as due to the same mechanism whereby singularities develop in the thermodynamic functions of systems experiencing phase transitions. Indeed we have discussed the fact that a singular fluctuation distribution function can be mapped onto a thermodynamic potential of a dual model with a critical point. The singularity appears similarly to what one observes in thermodynamic functions when a condensation transition is present. When such feature occurs at the level of fluctuations, at variance with the usual examples of condensation, one speaks of condensation of fluctuations.

Singularities of the probability distributions can have a scarce practical relevance if they occur in regions where fluctuations have a negligible chance to be observed. However, in some of the cases considered in this paper the non-analytical behavior is associated to the breakdown of the large deviation principle. As a result, large fluctuations of macrovariables have a better chance to be observed even in systems with a relatively large number of degrees of freedom. In this case the presence of singularities not only can be observed, but its effects can be appreciated. Perhaps, one of the most intriguing one is the so called giant susceptibility, whereby slightly tuning the properties of even one single component, say a molecule of a gas, can have catastrophic consequences on the behavior of the whole system.

Non-analyticity points in the probability distributions also influence the way in which rare fluctuations are developed out of typical state where they are absent. Indeed, it has been shown that large fluctuations in the region where condensation occurs are formed by means of a complex slow dynamics which resembles, once again a manifestation of a dual behavior, that of systems brought across a phase transition. The knowledge of the dynamical path leading to a rare fluctuation may have important consequences in those cases when such deviations lead to catastrophic events, as in the case of extinctions or bankruptcies.

Among the several perspectives of future studies in this context, we mention the possibility to explore the role of correlated noise on the large deviations, for instance in models of active particles where some analytical results can be obtained [66]; the meaning of singularities, which are related to non-equilibrium phase transitions, within the general framework of the macroscopic fluctuation theory [67]; the relation between the presence of singularities and the validity of the fluctuation relation for entropy production or related quantities in more general cases; the role of correlations among random variables in the anomalous large deviations, as observed for instance in conditioned random walks [68] and Brownian motion [69]; the effect of inhomogeneous rates in bulk-driven exclusion processes [70].

Author Contributions: F.C. and A.S. equally contributed to conceptualization, investigation and writing of the paper.

Funding: F.C. acknowledges funding from grant PRIN 2015K7KK8L covering also the publication costs of the present open access article. A.S. acknowledges support from "Programma VALERE" of University of Campania "L. Vanvitelli".

Acknowledgments: F.C. and A.S. acknowledge A. Crisanti, L.F. Cugliandolo, G. Gonnella, G. Gradenigo, A. Piscitelli, A. Puglisi, H. Touchette, A. Vulpiani and M. Zannetti for a long collaboration on some of the issues here discussed.

Conflicts of Interest: The authors declare no conflict of interest.

References

1. Langer, J.S. An introduction to the kinetics of first-order phase transitions. In *Solids Far from Equilibrium*; Godrèche, C., Ed.; Cambridge University Press: Cambridge, UK, 1991; pp. 297–363.
2. Merhav, N.; Kafri, Y. Bose–Einstein condensation in large deviations with applications to information systems. *J. Stat. Mech.* **2010**, *2010*, P02011. [CrossRef]
3. Filiasi, M.; Livan, G.; Marsili, M.; Peressi, M.; Vesselli, E.; Zarinelli, E. On the concentration of large deviations for fat tailed distributions, with application to financial data. *J. Stat. Mech.* **2014**, *2014*, P09030. [CrossRef]
4. Touchette, H. The large deviation approach to statistical mechanics. *Phys. Rep.* **2009**, *478*, 1–69. [CrossRef]
5. Vulpiani, A.; Cecconi, F.; Cencini, M.; Puglisi, A.; Vergni, D. From the Law of Large Numbers to Large Deviation Theory in Statistical Physics: An Introduction. In *Large Deviations in Physics*; Vulpiani, A., Cecconi, F., Cencini, M., Puglisi, A., Vergni, D., Eds.; Springer: Berlin/Heidelberg, Germany, 2014.
6. Patrick, A.E. Large deviations in the spherical model. In *On Three Levels*; Fannes, M., Maes, C., Verbeure, A., Eds.; Plenum Press: New York, NY, USA, 1994; p. 347.
7. Dobrushin, R.L.; Kotecky, R.; Shlosman, S. *Wulff Construction: A Global Shape from Local Interaction*; AMS Translation Series; AMS: Providence, RI, USA, 1992.
8. Pfister, C.E. Large deviations and phase separation in the two-dimensional Ising model. *Helv. Phys. Acta* **1991**, *64*, 953–1054.
9. Shlosman, S.B. The droplet in the tube: A case of phase transition in the canonical ensemble. *Commun. Math. Phys.* **1989**, *125*, 81–90. [CrossRef]
10. Harris, R.J.; Touchette, H.J. Current fluctuations in stochastic systems with long-range memory. *Phys. A Math. Theor.* **2009**, *42*, 342001. [CrossRef]
11. Gradenigo, G.; Sarracino, A.; Puglisi A.; Touchette, H. Fluctuation relations without uniform large deviations. *J. Phys. A Math. Theor.* **2013**, *46*, 335002. [CrossRef]
12. Gambassi A.; Silva, A. Large deviations and universality in quantum quenches. *Phys. Rev. Lett.* **2012**, *109*, 250602. [CrossRef]
13. den Hollander, F. *Large Deviations*; Fields Institute Monograph; AMS: Providence, RI, USA, 2000.
14. Corberi, F. Large deviations, condensation and giant response in a statistical system. *J. Phys. A Math. Theor.* **2015**, *48*, 465003. [CrossRef]
15. Bialas, P.; Burda, Z.; Johnston, D. Condensation in the backgammon model. *Nucl. Phys. B* **1997**, *493*, 505–516. [CrossRef]
16. Bialas, P.; Burda, Z.; Johnston, D. Phase diagram of the mean field model of simplicial gravity. *Nucl. Phys. B* **1999**, *542*, 413–424. [CrossRef]
17. Bialas, P.; Bogacz, L.; Burda, A.; Johnston, D. Finite size scaling of the balls in boxes model. *Nucl Phys. B* **2000**, *575*, 599–612. [CrossRef]
18. Evans M.R.; Hanney, T. Nonequilibrium statistical mechanics of the zero-range process and related models. *J. Phys. A Math. Gen.* **2005**, *38*, R195–R240. [CrossRef]
19. O'Loan, O.J.; Evans, M.R.; Cates, M.E. Jamming transition in a homogeneous one-dimensional system: The bus route model. *Phys. Rev. E* **1998**, *58*, 1404. [CrossRef]
20. Drouffe, J.-M.; Godrèche, C.; Camia, F. A simple stochastic model for the dynamics of condensation. *J. Phys. A Math. Gen.* **1998**, *31*, L19 [CrossRef]
21. Godrèche, C. From Urn Models to Zero-Range Processes: Statics and Dynamics. In *Ageing and the Glass Transition*; Lecture Notes in Physics; Henkel, M., Pleimling, M., Sanctuary, R., Eds.; Springer: Berlin, Germany, 2007; Volume 716.
22. Godrèche, C.; Luck, J.-M. Nonequilibrium dynamics of urn models. *J. Phys. Condens. Matter* **2002**, *14*, 1601. [CrossRef]
23. Godrèche, C.; Luck, J.-M. Nonequilibrium dynamics of the zeta urn model. *Eur. Phys. J. B* **2001**, *23*, 473. [CrossRef]

24. Gradenigo, G.; Bertin, E. Participation ratio for constraint-driven condensation with superextensive mass. *Entropy* **2017**, *19*, 517. [CrossRef]
25. Touchette, H.; Cohen, E.G.D. Fluctuation relation for a Lévy particle. *Phys. Rev. E* **2007**, *76*, 020101. [CrossRef]
26. Touchette, H.; Cohen, E.G.D. Anomalous fluctuation properties. *Phys. Rev. E* **2009**, *80*, 011114. [CrossRef]
27. Bouchet, F.; Touchette, H. Non-classical large deviations for a noisy system with non-isolated attractors. *J. Stat. Mech.* **2012**, *2012*, P05028. [CrossRef]
28. Harris, R.J.; Rákos, A.; Schütz, G.M. Current fluctuations in the zero-range process with open boundaries. *J. Stat. Mech.* **2005**, *2005*, P08003. [CrossRef]
29. Corberi, F.; Cugliandolo, L.F. Dynamic fluctuations in unfrustrated systems: Random walks, scalar fields and the Kosterlitz–Thouless phase. *J. Stat. Mech.* **2012**, *2012*, P11019. [CrossRef]
30. Szavits-Nossan, J.; Evans, M.R.; Majumdar, S.N. Constraint-Driven Condensation in Large Fluctuations of Linear Statistics. *Phys. Rev. Lett.* **2014**, *112*, 020602. [CrossRef]
31. Szavits-Nossan, J.; Evans, M.R.; Majumdar, S.N. Condensation transition in joint large deviations of linear statistics. *J. Phys. A Math. Theor.* **2014**, *47*, 455004. [CrossRef]
32. Chleboun, P.; Grosskinsky, S. Finite size effects and metastability in zero-range condensation. *J. Stat. Phys.* **2010**, *140*, 846–872. [CrossRef]
33. Zannetti, M. The grand canonical catastrophe as an instance of condensation of fluctuations. *Europhys. Lett.* **2015**, *111*, 20004. [CrossRef]
34. Corberi, F.; Gonnella, G.; Piscitelli, A. Singular behavior of fluctuations in a relaxation process. *J. Non-Cryst. Solids* **2015**, *407*, 51–56. [CrossRef]
35. Zannetti, M.; Corberi, F.; Gonnella, G. Condensation of fluctuations in and out of equilibrium. *Phys. Rev. E* **2014**, *90*, 012143. [CrossRef]
36. Zannetti, M.; Corberi, F.; Gonnella, G.; Piscitelli A. Energy and heat fluctuations in a temperature quench. *Commun. Theor. Phys.* **2014**, *62*, 555. [CrossRef]
37. Corberi, F.; Gonnella, G.; Piscitelli A.; Zannetti, M. Heat exchanges in a quenched ferromagnet. *J. Phys. A Math. Theor.* **2013**, *46*, 042001. [CrossRef]
38. Cagnetta, F.; Corberi, F.; Gonnella, G.; Suma, A. Large fluctuations and dynamic phase transition in a system of self-propelled particles. *Phys. Rev. Lett.* **2017**, *119*, 158002. [CrossRef]
39. Crisanti, A.; Sarracino, A.; Zannetti, M. Heat fluctuations of Brownian oscillators in nonstationary processes: Fluctuation theorem and condensation transition. *Phys. Rev. E* **2017**, *95*, 052138. [CrossRef] [PubMed]
40. Marini Bettolo Marconi, U.; Puglisi, A.; Rondoni, L.; Vulpiani, A. Fluctuation–dissipation: Response theory in statistical physics. *Phys. Rep.* **2008**, *461*, 111. [CrossRef]
41. Seifert, U. Stochastic thermodynamics, fluctuation theorems and molecular machines. *Rep. Prog. Phys.* **2012**, *75*, 126001. [CrossRef] [PubMed]
42. Baek, Y.; Kafri, Y. Singularities in large deviation functions. *J. Stat. Mech.* **2015**, *2015*, P08026. [CrossRef]
43. Corberi, F. Development and regression of a large fluctuation. *Phys. Rev. E* **2017**, *95*, 032136. [CrossRef]
44. Alastuey, A.; Piasecki, J. Approach to a stationary state in an external field. *J. Stat. Phys.* **2010**, *139*, 991–1012. [CrossRef]
45. Gradenigo, G.; Puglisi, A.; Sarracino, A.; Marini Bettolo Marconi, U. Nonequilibrium fluctuations in a driven stochastic Lorentz gas. *Phys. Rev. E* **2012**, *85*, 031112. [CrossRef]
46. Lebowitz, J.L.; Spohn, H. A Gallavotti–Cohen-type symmetry in the large deviation functional for stochastic dynamics. *J. Stat. Phys.* **1999**, *95*, 333–365. [CrossRef]
47. Puglisi, A.; Sarracino A.; Vulpiani, A. Temperature in and out of equilibrium: A review of concepts, tools and attempts. *Phys. Rep.* **2017**, *709–710*, 1–60. [CrossRef]
48. Gradenigo, G.; Majumdar, S.N. A First-Order Dynamical Transition in the displacement distribution of a Driven Run-and-Tumble Particle. *arXiv* **2018**, arXiv:1812.07819.
49. Mehl, J.; Speck, T.; Seifert, U. Large deviation function for entropy production in driven one-dimensional systems. *Phys. Rev. E* **2008**, *78*, 011123. [CrossRef] [PubMed]
50. Nyawo, P.T.; Touchette, H. Large deviations of the current for driven periodic diffusions. *Phys. Rev. E* **2016**, *94*, 032101. [CrossRef] [PubMed]
51. Speck, T.; Engel, A.; Seifert, U. The large deviation function for entropy production: The optimal trajectory and the role of fluctuations. *J. Stat. Mech.* **2012**, *2012*, P12001. [CrossRef]

52. Lacoste, D.; Lau, A.W.C.; Mallick, K. Fluctuation theorem and large deviation function for a solvable model of a molecular motor. *Phys. Rev. E* **2008**, *78*, 011915. [CrossRef]
53. Kumar, N.; Ramaswamy, S.; Sood, A.K. Symmetry properties of the large-deviation function of the velocity of a self-propelled polar particle. *Phys. Rev. Lett.* **2011**, *106*, 118001. [CrossRef] [PubMed]
54. Fischer, L.P.; Pietzonka, P.; Seifert, U. Large deviation function for a driven underdamped particle in a periodic potential. *Phys. Rev. E* **2018**, *97*, 022143. [CrossRef]
55. Budini, A.A. Fluctuation relations with intermittent non-Gaussian variables. *Phys. Rev. E* **2011**, *84*, 061118. [CrossRef]
56. Dorosz, S.; Pleimling, M. Entropy production in the nonequilibrium steady states of interacting many-body systems. *Phys. Rev. E* **2011**, *83*, 031107. [CrossRef]
57. Garrahan, J.P.; Jack, R.L.; Lecomte, V.; Pitard, E.; van Duijvendijk K.; van Wijland, F. First-order dynamical phase transition in models of glasses: An approach based on ensembles of histories. *J. Phys. A Math. Theor.* **2009**, *42*, 075007. [CrossRef]
58. Berlin, T.H.; Kac, M. The spherical model of a ferromagnet. *Phys. Rev.* **1952**, *86*, 821–835. [CrossRef]
59. Nagaev, S.V. Integral limit theorems taking large deviations into account when Cramér's condition does not hold, I. *Theory Probab. Appl.* **1969**, *14*, 51–64. [CrossRef]
60. Nagaev, S.V. Large deviations for sums of independent random variables. *Ann. Probab.* **1979**, *7*, 745–789. [CrossRef]
61. Chistyakov, V.P. A Theorem on Sums of Independent Positive Random Variables and Its Applications to Branching Random Processes. *Theory Probab. Appl.* **1964**, *9*, 640–648. [CrossRef]
62. Bray, A.J. Theory of phase-ordering kinetics. *Adv. Phys.* **1994**, *43*, 357–459. [CrossRef]
63. Onuki, A. *Phase Transition Dynamics*; Cambridge University Press: Cambridge, UK, 2004.
64. Puri, S.; Wadhawan, V. (Eds.) *Kinetics of Phase Transitions*; Taylor and Francis: London, UK, 2009.
65. Corberi, F. Coarsening in inhomogeneous systems. *Comptes Rendus Phys.* **2015**, *16*, 332–342. [CrossRef]
66. Marini Bettolo Marconi, U.; Puglisi, A.; Maggi, C. Heat, temperature and Clausius inequality in a model for active Brownian particles. *Sci. Rep.* **2017**, *7*, 46496. [CrossRef]
67. Bertini, L.; De Sole, A.; Gabrielli, D.; Jona-Lasinio, G.; Landim, C. Macroscopic fluctuation theory. *Rev. Mod. Phys.* **2015**, *87*, 593. [CrossRef]
68. Szavits-Nossan, J.; Evans, M.R.; Majumdar, S.N. Conditioned random walks and interaction-driven condensation. *J. Phys. A Math. Theor.* **2016**, *50*, 024005. [CrossRef]
69. Nickelsen, D.; Touchette, H. Anomalous Scaling of Dynamical Large Deviations. *Phys. Rev. Lett.* **2018**, *121*, 090602. [CrossRef]
70. Lazarescu, A. Generic dynamical phase transition in one-dimensional bulk-driven lattice gases with exclusion. *J. Phys. A Math. Theor.* **2017**, *50*, 254004. [CrossRef]

© 2019 by the authors. Licensee MDPI, Basel, Switzerland. This article is an open access article distributed under the terms and conditions of the Creative Commons Attribution (CC BY) license (http://creativecommons.org/licenses/by/4.0/).

Article

Daemonic Ergotropy: Generalised Measurements and Multipartite Settings

Fabian Bernards [1], Matthias Kleinmann [1], Otfried Gühne [1] and Mauro Paternostro [2,*]

[1] Naturwissenschaftlich-Technische Fakultät, Universität Siegen, Walter-Flex-Straße 3, 57068 Siegen, Germany
[2] School of Mathematics and Physics, Queen's University, Belfast BT7 1NN, UK
* Correspondence: m.paternostro@qub.ac.uk

Received: 6 July 2019; Accepted: 3 August 2019; Published: 7 August 2019

Abstract: Recently, the concept of daemonic ergotropy has been introduced to quantify the maximum energy that can be obtained from a quantum system through an ancilla-assisted work extraction protocol based on information gain via projective measurements [G. Francica et al., npj Quant. Inf. 3, 12 (2018)]. We prove that quantum correlations are not advantageous over classical correlations if projective measurements are considered. We go beyond the limitations of the original definition to include generalised measurements and provide an example in which this allows for a higher daemonic ergotropy. Moreover, we propose a see-saw algorithm to find a measurement that attains the maximum work extraction. Finally, we provide a multipartite generalisation of daemonic ergotropy that pinpoints the influence of multipartite quantum correlations, and study it for multipartite entangled and classical states.

Keywords: ergotropy; quantum correlations; information thermodynamics

1. Introduction

In the rapidly evolving research arena embodied by the thermodynamics of quantum systems, the resource-role of quantum features in work-extraction protocols is one of the most interesting and pressing open questions [1–4]. Quantum coherences are claimed to be responsible for the extraction of work from a single heat bath [5] and the enhanced performance of quantum engines [6]. Weakly driven quantum heat engines are known to exhibit enhanced power outputs with respect to their classical (stochastic) versions [7]. Quantum information-assisted schemes for energy extraction have been put forward and shown to be potentially able to achieve significant efficiencies [8–13]. However, controversies in the usefulness of quantum correlations and coherences in schemes for the extraction of work from quantum systems have also been discussed [14–17]. While a full physical understanding of these issues is still far from being acquired, theoretical progress in this direction will be key to the design and implementation of informed experimental proof-of-principle experiments and thus the consolidation of a quantum approach to the thermodynamics of microscopic systems.

Recently, a simple ancilla-assisted work-extraction protocol has been proposed that is able to pinpoint the crucial role that quantum measurements have in the performance of a quantum work-extraction game. This protocol also highlighted important implications arising from the availability of quantum correlations between the work medium and the ancilla [18]. The scheme provided a link between enhanced work extraction capabilities and quantum entanglement between ancilla and work medium, suggesting the possibility to exploit entanglement as a resource.

In this work we show that although this link exists for pure states, quantum correlations and work extraction capabilities are unrelated if mixed states are considered. However, the scheme in Reference [18] relied on a set of very stringent assumptions, which leave room to further investigations aimed at clarifying the potential benefits of exploiting quantum resources. Here, we critically investigate the protocol in Reference [18], and extend it in various directions. First, we address the class of measurements that ensure the enhancement of the work-extraction performance. We provide an example in which generalised measurements allow for more extracted energy than projective measurements do. The search for the right generalised measurement poses serious computational challenges that we solve by proposing a constructive see-saw algorithm that is able to identify the most effective measurement for a given state of the work medium and ancilla, and an assigned Hamiltonian of the former. We then address the issue embodied by the interplay between information gathered via optimal measurements and quantum correlations shared between work medium and ancilla. We show that, depending on the nature of the optimal measurement, quantum correlations may become entirely inessential for the enhancement of work extraction. Finally, we open the investigation to multipartite settings by addressing the case of multiple work media and ancillas, showing that the structure of correlation-sharing among the various parties of such a system is key in the performance of our work-extraction protocol.

Our results contribute to the ongoing research for the ultimate resources to be exploited to draw an effective and useful framework for quantum enhanced thermodynamical processes. While clarifying a number of important points, our work opens up new avenues of investigation that will be crucial for the design of unambiguous experimental validations.

2. Notation and Concepts

The maximal energy decrease of a given state ϱ^S with respect to a reference Hamiltonian H undergoing an arbitrary unitary evolution U is its ergotropy [19]

$$W(\varrho^S, H) = \mathrm{Tr}[\varrho^S H] - \min_U \mathrm{Tr}[U\varrho^S U^\dagger H]. \tag{1}$$

This is interpreted as the maximal amount of work that can be extracted from a system prepared in state ϱ^S by the means of a unitary protocol [19]. Given some state in its spectral decomposition $\varrho^S = \sum_k r_k |r_k\rangle\langle r_k|$ with $r_{k+1} \leq r_k$ and a Hamiltonian $H = \sum_k \epsilon_k |\epsilon_k\rangle\langle \epsilon_k|$ with $\epsilon_{k+1} \geq \epsilon_k$ the optimal unitary is $U = \sum_k |\epsilon_k\rangle\langle r_k|$ [19]. This is a direct consequence of the von Neumann trace inequality [20]. It states that $\mathrm{tr}[AB] \leq \sum_i a_i b_i$, where a_i (b_i) are the eigenvalues of A (B) and $a_{i+1} \geq a_i$, $b_{i+1} \geq b_i$. Choosing $A = -U\varrho^S U\dagger$ and $B = H$ and writing $\max_U \mathrm{tr}[-U\varrho^S U + H] = -\min_U \mathrm{tr}[U\varrho^S U + H]$ then shows that the bound given by the von Neumann trace inequality is achieved with the unitary stated above.

In Reference [18], an ancilla-assisted protocol allowed for enhanced work extraction by making use of a process of information inference. The fundamental building blocks of the protocol are embodied by the joint state of a work medium S and an ancilla A, and a projective measurement M performed on the latter (cf. Figure 1). The information gathered through these measurements is then used to determine a unitary transformation to be applied to S to extract as much work as possible.

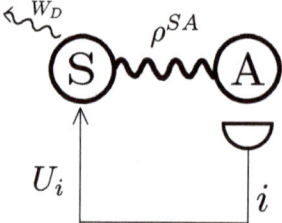

Figure 1. Illustration of daemonic ergotropy. A system S is coupled to an ancilla A. A measurement is performed on the latter and depending on the outcome i different unitaries can be applied to S in order to extract work. The maximal amount of extractable work using this protocol is the daemonic ergotropy.

This work, which is dubbed daemonic ergotropy, is given by

$$W_D(\varrho^{SA}, H, M) = \text{Tr}[\varrho^S H] - \sum_i \min_{U_i} \text{Tr}(\gamma_i^S \tilde{H}_i), \quad (2)$$

where $\tilde{H}_i = U_i^\dagger H U_i$, $M = \{\Pi_j\}$ is a projective measurement, and $\gamma_i^S = \text{Tr}_A[\varrho^{SA}(I^S \otimes \Pi_i^A)]$ is the unnormalised conditional state of S corresponding to the ith measurement outcome. The daemonic ergotropy can be written in a more compact way using the ergotropy, namely

$$W_D(\varrho^{SA}, H, M) = \sum_i W(\gamma_i^S, H). \quad (3)$$

For a pure state, any projective measurement M with Π_i rank-one projectors maximises the daemonic ergotropy. In fact, the conditional states γ_i^S are then pure and it is always possible to find a unitary—specific to every conditional state—that maps it to the ground state of the Hamiltonian, thus lowering as much as possible the energy of the system and extracting the maximum amount of work [18].

The difference between maximal daemonic ergotropy and ergotropy is called daemonic gain [18], and is formalised as

$$\delta W(\varrho^{SA}, H) = \max_M W_D(\varrho^{SA}, H, M) - W(\varrho^S, H). \quad (4)$$

If ϱ^{SA} is a pure product state, ϱ^S is pure. Thus, no measurement on the ancilla is required for optimal work extraction, since in this case there is a unitary that maps ϱ^S to the ground state of the Hamiltonian. Consequently, the daemonic ergotropy coincides with the ergotropy in this case and there is no daemonic gain.

The definitions provided above pinpoint the key role of the measurement step in such an ancilla-assisted extraction protocol. In particular, the assumption of projective measurements performed on A appears to be too restrictive. It is thus plausible to wonder if better performances of the daemonic work-extraction scheme are possible when enlarging the range of possible measurements on the ancilla to generalised quantum measurements.

3. Non-Optimality of Projective Measurements for Daemonic Ergotropy

We now address such a scenario and provide an example where more energy can be extracted from S when generalised measurements are performed. To this end, we will employ the formalism of positive operator valued measures (POVMs) [21]. In the case of a finite set of outcomes $\{i\}$, a POVM is a map that assigns a positive semidefinite operator E_i—dubbed as effect—to each outcome i, such that $\sum_i E_i = I$. As with projective measurements, the probabilities for the outcomes are obtained as $p_i = \text{Tr}(E_i \varrho)$. However, the effects E_i of a POVM need not be projectors.

Let us consider now a three-level system S and a two-level ancilla A prepared in the joint state

$$\varrho^{SA} = \frac{1}{3} \sum_{j=0}^{2} |j\rangle\langle j| \otimes \Pi\left(\frac{2\pi j}{3}, 0\right) \quad (5)$$

with projectors

$$\Pi(\alpha, \beta) = \frac{1}{2}\{I + \cos(\alpha)\sigma_z + \sin(\alpha)[\cos(\beta)\sigma_x - \sin(\beta)\sigma_y]\}. \quad (6)$$

Here (α, β) are angles in the single-qubit Bloch sphere. We assume a reference Hamiltonian $H = \sum_j \epsilon_j |j\rangle\langle j|$ with energy eigenvalues ϵ_j arranged in increasing order. If only projective measurements M are allowed on the state of the ancilla, the maximum daemonic ergotropy achieved upon optimizing over the measurement strategy is

$$\max_{M} W_D(\varrho^{SA}, H, M) = W(\varrho^S, H) + \frac{\epsilon_2 - \epsilon_0}{2\sqrt{3}}. \tag{7}$$

Details on this result are presented in Appendix A. However, if generalised measurements are permitted, one may choose the POVM with effects $E_j = \frac{2}{3}\Pi(2\pi j/3, 0)$ to yield a daemonic ergotropy of

$$W_D(\varrho^{SA}, H, \{E_i\}) = W(\varrho^S, H) + \frac{1}{6}(\epsilon_1 + \epsilon_2 - 2\epsilon_0). \tag{8}$$

This can exceed the maximum daemonic ergotropy achieved through projective measurements. For instance, we can assume to have shifted energy so that $\epsilon_0 = 0$. Under such conditions, we would have $W_D(\varrho^{SA}, H, \{E_i\}) > \max_M W_D(\varrho^{SA}, H, M)$ for $(\sqrt{3}-1)\epsilon_2 < \epsilon_1 \leq \epsilon_2$. Figure 2 shows the daemonic gain δW corresponding to the example above as a function of the value of the highest energy level of the Hamiltonian for projective measurements (PVMs) and POVMs. While in this example the optimal projective measurement does not depend on the Hamiltonian, the optimal POVM does. Therefore, the daemonic gain grows linearly with the value of the highest energy value, as long as only projective measurements are taken into account. For comparison, the daemonic gain that can be achieved with the previously discussed POVM $\left(\frac{2}{3}\Pi(2\pi j/3, 0)\right)_j$ is plotted as a dashed line.

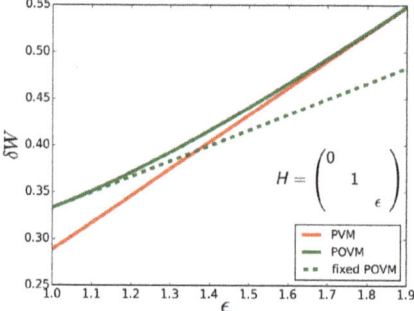

Figure 2. Daemonic gain δW as a function of the value of the highest energy level of the Hamiltonian H (in units of ϵ_1) for the state ϱ^{SA} given in Equation (5). Here $\epsilon = \epsilon_2/\epsilon_1$. We compare the performance under the optimal r projective measurements (PVM) and positive operator valued measures (POVM). The latter was found numerically using the see-saw algorithm proposed here. The former is determined analytically as discussed in Appendix A. The dashed line is obtained as the daemonic gain δW for the fixed POVM with effects $E_j = \frac{2}{3}\Pi(2\pi j/3, 0)$.

4. Construction of Optimal POVMs

Having provided a useful example, we now move to address the problem of identifying the ideal POVM for optimal daemonic ergotropy. The following Lemma is instrumental to the achievement of our goal:

Lemma 1. *The ergotropy is a sublinear function in its first argument, which refers to the state. That is, for any $\gamma = \gamma_1 + \gamma_2$*

$$W(\gamma, H) \leq \sum_{i=1,2} W(\gamma_i, H) \tag{9}$$

and

$$W(\lambda \gamma, H) = \lambda W(\gamma, H) \tag{10}$$

for any $\lambda \geq 0$. As ergotropy is symmetric under the exchange of its first and the second argument, it is also sublinear in the Hamiltonian.

Proof. The second equation holds trivially, which justifies our use of unnormalised states. We obtain the first inequality as follows

$$\begin{aligned}
W(\gamma, H) &= \text{Tr}(\gamma H) - \min_U \text{Tr}[U\gamma U^\dagger H] \\
&\leq \sum_{j=1,2} \left[\text{Tr}(\gamma_j H) - \min_U \text{Tr}(U\gamma_j U^\dagger H) \right] \\
&= \sum_{j=1,2} W(\gamma_j, H).
\end{aligned} \qquad (11)$$

□

Note that sublinearity implies convexity, i.e., $W[\lambda\gamma_1 + (1-\lambda)\gamma_2, H] \leq \lambda W(\gamma_1, H) + (1-\lambda)W(\gamma_2, H)$. This result allows us to state the following corollary:

Corollary 2. *The daemonic ergotropy*

$$W_D(\varrho^{SA}, H, M) = \sum_i W(\gamma_i^S, H) \geq W(\sum_i \gamma_i^S, H) = W(\varrho^S, H) \qquad (12)$$

is larger or equal to ergotropy. Equality holds for the trivial measurement, with the identity as only effect.

This claim has already been proven in a different way in Reference [18]. A second interesting consequence of the sublinearity of ergotropy is stated in the following lemma:

Lemma 3. *Daemonic ergotropy is a convex function of its third argument, which pertains to the measurement strategy.*

Proof. Let us consider a mixed measurement strategy $Q = \lambda M + (1-\lambda)N$ with $0 \leq \lambda \leq 1$, and the corresponding daemonic ergotropy. We have

$$\begin{aligned}
W_D[\varrho^{SA}, H, Q] &\leq \lambda \sum_i W[\text{Tr}_A(\varrho^{SA} I \otimes M_i), H] + (1-\lambda) \sum_i W[\text{Tr}_A(\varrho^{SA} I \otimes N_i), H] \\
&= \lambda W_D(\varrho^{SA}, H, M) + (1-\lambda)W_D(\varrho^{SA}, H, N).
\end{aligned} \qquad (13)$$

□

We complete our formal analysis that precedes the presentation of an algorithm for the identification of the optimal POVM with the following theorem.

Theorem 4. *For any state ϱ^{SA} and any POVM M, one can find a POVM $\widetilde{M}$ with at most d^2 effects, where d is the dimension of the ancilla, such that*

$$W_D(\varrho^{SA}, H, M) - W_D(\varrho^{SA}, H, \widetilde{M}). \qquad (14)$$

Proof. The set of POVMs on a d dimensional system is convex and it has been shown that the extremal points of this set are POVMs with at most d^2 effects [22]. A convex function that is defined on a convex domain takes its maximum on an extremal point. Therefore, there is an extremal POVM E with n outcomes, $1 \leq n \leq d^2$, that exhibits a daemonic ergotropy that is larger than or equal to the daemonic ergotropy for M. If equality holds, we choose $\widetilde{M} = E$. Otherwise, there is a mixture $\widetilde{M} = \lambda E + (1-\lambda)I$

between E and a trivial random measurement I with n outcomes and effects $I_i = I/n$ that meets the requirement, since $W_D(\varrho^{SA}, H, I) = W(\varrho^S, H) \leq W(\varrho^{SA}, H, M)$. □

We are now in the position to present an algorithm for the search of the optimal measurement. This task involves two parts (a) Finding the optimal measurement and (b) Finding the optimal unitaries to calculate the ergotropies of the conditional states. Assume a fixed measurement. Then, the conditional states are fixed and one can find the optimal unitaries as discussed in the introduction after Equation (1). On the other hand, if some d^2 unitaries U_i are given, finding the optimal measurement $M = (E_i)_i$ is a semidefinite program (SDP) [23]

$$\min_M \sum_i \mathrm{Tr}(\tau_i E_i)$$
$$\text{s.t} \sum_i E_i = I$$
$$E_i \geq 0 \qquad (15)$$

where E_i are the effects associated with the POVM M and

$$\tau_i = \mathrm{Tr}_S(\varrho^{SA} U_i^\dagger H U_i). \qquad (16)$$

We thus propose the following see-saw Algorithm 1:

Algorithm 1 Optimise POVM for daemonic ergotropy

1: Choose n different unitaries U_i and calculate τ_i
2: Solve the SDP above. This will yield a POVM M.
3: Calculate the conditional states γ_i^S for the POVM M and the optimal unitaries U_i.
4: **repeat** ▷ Iterate steps 2 and 3
5: **until** convergence.

We can restrict ourselves to $n = d^2$ different unitaries in the first step because of Theorem 4. Calculating the daemonic ergotropy after every round of the algorithm will yield a monotonically increasing sequence that is bounded from above because all involved operators are bounded and will therefore converge. In the case of the example discussed above, roughly 10 iterations are needed until the limit is reached within numerical precision. The sequence however sometimes converges to a local maximum that is strictly smaller than the maximal daemonic ergotropy. Besides observing this in practice, we also construct such a case in Appendix B.

5. The Role of Quantum Correlations

Notwithstanding the handiness of the algorithm built above, analytical solutions can be found in some physically relevant cases. The one most pertinent to the scopes of this work [18] is embodied by quantum-classical S-A states, i.e., states that can be cast in the form

$$\varrho_{qc}^{SA} = \sum_j \sigma_j^S \otimes |j\rangle\langle j|^A \qquad (17)$$

with $\{|j\rangle^A\}$ a set of orthonormal vectors and σ_j^S unnormalised states. This class of states has attracted attention from the community interested in the characterization of general quantum correlations, for it has only classical correlations, that is, it is not entangled and exhibits no quantum discord, if A is considered as the system the measurement being performed on [24–27]. For these states, we provide the following theorem. The proof is found in Appendix C.

Theorem 5. For a quantum-classical state ϱ_{qc}^{SA}, the maximum daemonic ergotropy is

$$\max_M W_D(\varrho^{SA}, H, M) = \sum_j W(\sigma_j^S, H). \qquad (18)$$

This value is achieved by performing the projective measurement with effects $P_j = |j\rangle\langle j|^A$ ($j = 1, \ldots, d$) on the ancilla A.

This shows that, in the case of a quantum-classical state, we have an analytic form for the daemonic gain. To calculate it, we should diagonalise the reduced state $\varrho^A = \text{Tr}_S(\varrho^{SA})$ of the ancilla. This yields a unitary to make the state block-diagonal. The individual blocks are then the optimal conditional states σ_j^S that one needs in order to compute the daemonic gain.

The above result paves the way to an investigation on the role that quantum correlations play in the daemonic protocol for work extraction. This important question was already partially addressed in Reference [18], where a very close relation between daemonic gain and entanglement in pure S-A states was pointed out, while the link was shown to be looser for the case of mixed resource states.

Here, by using the results reported above, we shed further light on the link between daemonic gain and quantum correlations. Let us assume that, for a given resource state ϱ^{SA}, the optimal measurement for daemonic gain is projective, and call $P_i = |i\rangle\langle i|$ the corresponding projections, which can be chosen, without loss of generality, to be rank one. We write the resource state as

$$\varrho^{SA} = \sum_{ij}^S \sigma_{ij}^S \otimes |i\rangle\langle j|^A, \qquad (19)$$

where the dyads $|i\rangle\langle j|^A$ are written in the basis defined by the optimal projectors P_i above. We notice that all off-block-diagonal terms σ_{ij}^S (with $i \neq j$) do not contribute to the daemonic gain, which is thus the same as the one associated with the quantum-classical state

$$\varrho_{qc}^{SA} = \sum_i \sigma_{ii}^S \otimes |i\rangle\langle i|^A. \qquad (20)$$

That this state is a quantum-classical state is obvious from the definition provided in Equation (17). This state can be produced by performing the optimal measurement and preparing a pure state on the ancilla accordingly. This procedure destroys all the quantum correlations, while the daemonic gain remains unchanged. Quantum correlations in the resource states are thus not useful, if the optimal measurement is projective. This is especially true if only projective measurements are considered from the start, which stresses the importance of considering generalised measurements, if one aims at investigating the impact entanglement may have on daemonic ergotropy.

However, we now show that, even if we allow for the use of arbitrary POVMs, the maximum daemonic gain for any given Hamiltonian can be achieved by classical-classical states, i.e., states whose parties share only classical correlations [26]. We do this by providing an upper bound on the daemonic gain. This bound is tight as it is achieved by a classical-classical state. Let us consider an explicit formula for daemonic gain, where we have inserted the definitions of ergotropy and daemonic ergotropy. We have

$$\delta W(\varrho^{SA}, H) = \min_U \text{Tr}(U\varrho^S U^\dagger H) - \min_{(E_k)} \min_{U_k} \sum_k \text{Tr}(U_k \varrho_k^S U_k^\dagger H). \qquad (21)$$

Using von Neumann's trace inequality, which reads $\text{Tr}(AB) \leq \sum_i a_i b_i$ with $a_i(b_i)$ the eigenvalues of A (B) in increasing order, one easily finds that the first term never exceeds $\frac{1}{d_S}\text{Tr}(H)$, where d_S is the dimension of the Hilbert space of S. This value is attained if ϱ^S is maximally mixed. The smallest value that the second term can take is ϵ_0, the lowest energy eigenvalue. This is achieved for pure conditional states ϱ_k^S. Consequently

$$\delta W(\varrho^{SA}, H) \leq \frac{1}{d_S}\text{Tr}(H) - \epsilon_0. \tag{22}$$

If the dimension of the ancilla d_A is greater or equal to d_S, this value is attained by using—among others—the classical-classical state

$$\varrho^{SA} = \frac{1}{d_S}\sum_{i=1}^{d_S}|s_i\rangle\langle s_i|^S \otimes |a_i\rangle\langle a_i|^A \tag{23}$$

and the projective measurement with effects $|a_i\rangle\langle a_i|^A$, where $\{|a_i\rangle^A\}$ ($\{|s_i\rangle^S\}$) forms an orthogonal basis of A (S). In the above example, the bound is also achievable with maximally entangled pure states

$$|\Psi^{SA}\rangle = \frac{1}{\sqrt{d_S}}\sum_{i=1}^{d_S}|s_i\rangle^S|a_i\rangle^A. \tag{24}$$

The maximal daemonic gain is, however, not always achieved using pure states, as the following example shows. Consider the following classical-classical state with a qutrit system and a qubit ancilla

$$\varrho^{SA} = \frac{1}{3}[|0\rangle\langle 0|^S \otimes |0\rangle\langle 0|^A + (|1\rangle\langle 1|^S + |2\rangle\langle 2|^S) \otimes |1\rangle\langle 1|^A]. \tag{25}$$

For a Hamiltonian with eigenvalues $\epsilon_0 \leq \epsilon_1 \leq \epsilon_2$ one easily finds the daemonic gain $\delta W(\varrho) = (\epsilon_2 - \epsilon_0)/3$. On the other hand, for any pure state, including maximally entangled states, we have

$$\delta W(|\Psi\rangle^{SA}) \leq \frac{1}{2}(\epsilon_1 - \epsilon_0), \tag{26}$$

since the Schmidt-rank of a pure state on a 3×2 dimensional system is at most 2. For a suitably chosen Hamiltonian, such as $H/\epsilon_1 = |1\rangle\langle 1| + \epsilon|2\rangle\langle 2|$, with $\epsilon = \epsilon_2/\epsilon_1 > 3/2$, the daemonic gain of ϱ^{SA} [Equation (25)] exceeds the daemonic gain of any pure state of the same system.

6. Multipartite Daemonic Ergotropy

In this section we want to investigate a multipartite adaptation of the daemonic ergotropy protocol. Concretely, we consider the situation in which N different parties $i \in \{1, ..., N\}$ each own one system S_i, whose energy they can locally measure using their local Hamiltonian $H^{(i)}$. The energy of all systems combined will then be evaluated using the Hamiltonian

$$H = \sum_{i=1}^{N} H^{(i)}. \tag{27}$$

Additionally, they can only act on their systems locally, that is using local unitaries. It is only this restriction that makes the protocol multipartite regarding the systems. If arbitrary global unitaries were admitted, this would be equivalent to a situation with a single system consisting of N subsystems.

We also take the case into account in which there are M ancillas, each owned by a different party $k \in \{1, ..., M\}$. As we are interested in a genuinely multipartite protocol, each party must resort to local measurements, possibly assisted by classical communication among the parties, yielding outcomes j_k. After all outcomes are obtained, they are publicly announced and every party i performs a unitary on their system S_i, which may depend on all the outcomes $\vec{j} = (j_k)_{k=1}^{M}$. We define the multipartite daemonic ergotropy W_D^{mult} to be the maximum amount of energy that can be extracted from a state in this way.

Note that, in spite of the previously imposed restrictions, our notion of multipartite daemonic ergotropy is in fact a generalisation of daemonic ergotropy. This might appear paradoxical at first glance. However, the daemonic ergotropy protocol is equivalent to the protocol of multipartite daemonic ergotropy for one system and one ancilla. This especially includes scenarios in which system and ancilla comprise several subsystems. Studying multipartite daemonic ergotropy is interesting, because it is also applicable to settings, in which the implementation of global measurements and unitaries are unfeasible.

As we are only concerned with local measurements, possibly assisted by classical communication among the parties, all effects of a POVM are of the form

$$E_{\vec{j}} = \bigotimes_{k=1}^{M} E_{j_k}^k. \tag{28}$$

We denote the respective conditional states of all systems by $\varrho_{\vec{j}}^S = \mathrm{Tr}_{(A_1...A_M)}(\varrho^{S_1...S_N A_1...A_M} E_{\vec{j}})$ and the conditional state of system S_i given a measurement outcome $\vec{j}$ as $\varrho_{\vec{j}}^i$. As before, the multipartite daemonic ergotropy can then be expressed in terms of the ergotropy as

$$W_D^{\mathrm{mult}}(\varrho^{\{S_i\},\{A_k\}}, H, E) = \sum_{\vec{j}} \sum_{i=1}^{N} W(\varrho_{\vec{j}}^i, H^{(i)}). \tag{29}$$

With this result, we can show that contrary to the bipartite case [cf. discussions after Equation (3)] in the multipartite setting projective measurements are in general not optimal for work extraction even for pure states. In order to see this, consider a state $\varrho^{S_1 A}$ and a purification $|\psi\rangle^{S_1 S_2 A}$, with $\varrho^{S_1 A} = \mathrm{Tr}_{S_2}(|\psi\rangle\langle\psi|^{S_1 S_2 A})$. If we now assume that system S_2 is equipped with a local Hamiltonian $H^{(2)} = hI$, where h is a constant, the multipartite daemonic ergotropy of the purified state is

$$W_D^{\mathrm{mult}}(|\psi\rangle^{S_1 S_2 A}, H, E) = \sum_{\vec{j}} \left[W(\varrho_{\vec{j}}^1, H^{(1)}) + W(\varrho_{\vec{j}}^2, H^{(2)}) \right]$$

$$= \sum_{\vec{j}} W(\varrho_{\vec{j}}^1, H^{(1)}) \tag{30}$$

$$= W_D(\varrho^{S_1 A}, H^{(1)}, E).$$

This result stems from the fact that $H^{(2)}$ is completely degenerate and the ergotropy vanishes for such Hamiltonians. Thus, also the multipartite daemonic ergotropy of the purification is maximised for the same POVM that also maximises the daemonic ergotropy of ϱ^{SA}. Hence, the purification of the qutrit-qubit state stated in Equation (5) is an example for a pure state that requires a POVM to maximise the multipartite daemonic ergotropy. Note, however, that there are also states for which projective measurements are optimal independently of the choice of the Hamiltonian. The first example are states that possess a Schmidt decomposition [28], i.e.,

$$|\Psi\rangle = \sum_i \sqrt{\lambda_i} |i_{S_1} \ldots i_{S_n} i_{A_1} \ldots i_{A_m}\rangle, \tag{31}$$

with $\langle i_{S_l} | j_{S_l}\rangle = \langle i_{A_l} | j_{A_l}\rangle = \delta_{ij} \forall i, j, l$. For qubits, these are exactly the states that become separable as soon as one particle is ignored [29]. A famous example is the m-partite Greenberger–Horne–Zeilinger (GHZ) state

$$|GHZ\rangle = \frac{1}{\sqrt{2}}(|0_{S_1} \ldots 0_{S_n} 0_{A_1} \ldots 0_{A_m}\rangle + |1_{S_1} \ldots 1_{S_n} 1_{A_1} \ldots 1_{A_m}\rangle), \tag{32}$$

for which the local projective measurements on $|0\rangle$ and $|1\rangle$ are optimal, since the conditional state of all systems is a pure product state independently of the outcome and its energy can thus be minimised using local unitaries.

A second class of states for which projective measurements are always optimal are multipartite quantum-classical states

$$\varrho_{S_1...S_nA} = \sum_i \sigma_i^{S_1...S_n} \otimes |i\rangle\langle i|^A. \tag{33}$$

Here, we can recover the proof of Theorem 5 to show that the projective measurement with projectors $|i\rangle\langle i|$ is optimal. The only adaptation to the proof is that the unitaries are now required to be products. Of course this result is still true in the special case when the ancilla is made up of several parties, such that the state can be written as

$$\varrho^{\{S_j\}...\{A_m\}} = \sum_i \sigma_i^{S_1...S_n} \otimes |i\rangle\langle i|^{A_1} \otimes \ldots |i\rangle\langle i|^{A_m}. \tag{34}$$

In this case, the optimal measurement consists of the local projective measurements with effects $|i\rangle\langle i|_{A_k}$.

7. Conclusions

We have significantly extended the concept of daemonic ergotropy to situations involving POVM-based information-gain processes, demonstrating that, in general, one should expect an advantage coming from the use of generalised quantum measurements in ancilla-assisted work-extraction schemes. While the optimal generalised measurements can be identified analytically in some restricted—yet physically relevant—cases, we have proposed an SDP-based see-saw algorithm for their construction. This has led to a number of results shedding light on previously unreported issues linked to daemonic approaches to quantum work extraction: while the interplay between quantum correlations and the features of the optimal measurements appears to be intricate, the structure of entanglement sharing in a multipartite scenario where only local unitaries and POVMs are used turns out to be key in the performance of ancilla-assisted work extraction.

Our work paves the way to a number of interesting developments aimed at exploring further and clarifying the relation between quantum features and work-extraction games in quantum scenarios. On the one hand, it will be very interesting to further compare, quantitatively, the performance of daemonic protocols under optimal PVMs and POVMs to ascertain the extents of the benefits induced by the latter class of measurements against the difficulty of practically implement them. On the other hand, the analysis that we have reported here leaves room to the in-depth assessment of multipartite daemonic gain against the structure of multipartite entanglement aimed at the identification of potentially *optimal* classes of multipartite entangled states, when gauged against their role as a resource in work-extraction schemes.

Author Contributions: M.P. suggested the original problem to tackle; F.B. M.K., O.G. and M.P. identified the methodology and technical tools; F.B. developed the project and wrote the first draft of the manuscript, which was then finalized by all the authors.

Funding: M.P. acknowledges support by the SFI-DfE Investigator Programme (grant 15/IA/2864), the H2020 Collaborative Project TEQ (Grant Agreement 766900), the Leverhulme Trust Research Project Grant UltraQuTe (grant nr. RGP-2018-266) and the Royal Society Wolfson Fellowship (RSWF\R3\183013). O.G. acknowledges support by the DFG and the ERC (Consolidator Grant 683107/TempQ).

Acknowledgments: F.B. thanks the Centre for Theoretical Atomic, Molecular and Optical Physics for hospitality while developing on part of this work as well as the Studienstiftung des Deutschen Volkes e.V. and the House of Young Talents Siegen.

Conflicts of Interest: The authors declare no conflict of interest.

Appendix A. POVM Advantage in Qutrit-Qubit Example

We present the state

$$\varrho^{SA} = \frac{1}{3}\sum_{j=0}^{2}|j\rangle\langle j| \otimes P_j \quad (A1)$$

with

$$P_j = \Pi\left(\frac{2\pi j}{3}, 0\right) \quad (A2)$$

and

$$\Pi(\alpha, \beta) = \frac{1}{2}\{I + \cos(\alpha)\sigma_z + \sin(\alpha)[\cos(\beta)\sigma_x - \sin(\beta)\sigma_y]\} \quad (A3)$$

as an example in which higher daemonic ergotropy can be achieved with POVMs compared to projective measurements, if a Hamiltonian is chosen suitably. Here, we work out the details and show all necessary calculations explicitly. First, we find the optimal projective measurements. It turns out, that they can be found independently of the chosen Hamiltonian. With this result and bearing in mind that the daemonic gain is invariant under unitary transformations of the Hamiltonian, we can then compute the daemonic ergotropy as a function of the energy spectrum.

Since the ancilla is a qubit, there are only two types of projective measurements: Either, the projective measurement has one outcome that is obtained with certainty, which makes the measurement trivial, or the measurement has two outcomes. In the latter case, the effects are rank one. Therefore, we can compute the maximal daemonic gain for projective measurements by computing it for the measurement $\Pi = (\Pi(\alpha, \beta), \Pi(\alpha + \pi, \beta))$ and optimise over the angles α and β afterwards. We have

$$\varrho^S = \frac{1}{3}(|0\rangle\langle 0| + |1\rangle\langle 1| + |2\rangle\langle 2|),$$

$$\varrho_\alpha^S = \text{Tr}[\varrho^{SA}(I \otimes \Pi(\alpha, \beta))]$$

$$= \frac{1}{3}\{|0\rangle\langle 0|\text{Tr}[P_0\Pi(\alpha, \beta)] + |1\rangle\langle 1|\text{Tr}[P_1\Pi(\alpha, \beta)] + |2\rangle\langle 2|\text{Tr}[P_2\Pi(\alpha, \beta)]\}$$

$$= \frac{1}{3}\left[|0\rangle\langle 0|\frac{1}{2}(1+\cos(\alpha)) + |1\rangle\langle 1|\left(\frac{1}{2} - \frac{1}{4}\cos(\alpha) + \frac{\sqrt{3}}{4}\sin(\alpha)\cos(\beta)\right)\right.$$

$$\left. + |2\rangle\langle 2|\left(\frac{1}{2} - \frac{1}{4}\cos(\alpha) - \frac{\sqrt{3}}{4}\sin(\alpha)\cos(\beta)\right)\right],$$

$$\varrho_{\alpha+\pi}^S = \text{Tr}[\varrho_{SA}(I \otimes \Pi(\alpha+\pi, \beta))]$$

$$= \frac{1}{3}\left[|0\rangle\langle 0|\frac{1}{2}(1-\cos(\alpha)) + |1\rangle\langle 1|\left(\frac{1}{2} + \frac{1}{4}\cos(\alpha) - \frac{\sqrt{3}}{4}\sin(\alpha)\cos(\beta)\right)\right.$$

$$\left. + |2\rangle\langle 2|\left(\frac{1}{2} + \frac{1}{4}\cos(\alpha) + \frac{\sqrt{3}}{4}\sin(\alpha)\cos(\beta)\right)\right]. \quad (A4)$$

From the definition of ergotropy one can easily see that the ergotropy of the conditional states γ_α^S and $\gamma_{\alpha+\pi}^S$ will be maximal for $\cos(\beta) = \pm 1$. This becomes clear when considering a state

$$\varrho = a|0\rangle\langle 0| + (b+c)|1\rangle\langle 1| + (b-c)|2\rangle\langle 2|, \quad (A5)$$

where $a, b, c \in \mathbb{R}$ and $c \geq 0$. Let the Hamiltonian be

$$H = \epsilon_0 |\epsilon_0\rangle\langle\epsilon_0| + \epsilon_1 |\epsilon_1\rangle\langle\epsilon_1| + \epsilon_2 |\epsilon_2\rangle\langle\epsilon_2|. \tag{A6}$$

Then, the ergotropy can without loss of generality be written as

$$\begin{aligned} W &= \text{Tr}[\varrho H] - \min_U \text{Tr}[U\varrho U^\dagger H] \\ &= \text{Tr}[\varrho H] - [\epsilon_0 a + \epsilon_1(b+c) + \epsilon_2(b-c)] \\ &= \text{Tr}[\varrho H] - [a\epsilon_0 + b(\epsilon_1 + \epsilon_2) + c(\epsilon_1 - \epsilon_2)], \end{aligned} \tag{A7}$$

where the energy eigenvalues are ordered such that the minimum is achieved. Consequently, we get $\epsilon_1 \leq \epsilon_2$ since $(b+c) \geq (b-c)$. Therefore, W increases with c and we can set $\beta = 0$ in the above calculation. Exploiting addition theorems, we can now write

$$\varrho_\alpha^S = \frac{1}{6}\left[|0\rangle\langle 0|[1 + \cos(\alpha)] + |1\rangle\langle 1|\left(1 + \cos\left(\alpha - \frac{2\pi}{3}\right)\right) + |2\rangle\langle 2|\left(1 + \cos\left(\alpha + \frac{2\pi}{3}\right)\right)\right]$$

$$\varrho_{\alpha+\pi}^S = \frac{1}{6}\left[|0\rangle\langle 0|(1 - \cos\alpha) + |1\rangle\langle 1|\left(1 - \cos\left(\alpha - \frac{2\pi}{3}\right)\right) + |2\rangle\langle 2|\left(1 - \cos\left(\alpha + \frac{2\pi}{3}\right)\right)\right].$$

As one can easily see, an optimal value of α is not unique, as shifting its value by $\frac{2\pi}{3}$ can be compensated by relabeling the states, which does not affect the daemonic gain. We now aim to find the optimal α in the interval $[-\frac{\pi}{3}, \frac{\pi}{3})$. When calculating the ergotropy of the conditional states we need to know the ordering of their eigenvalues

$$\alpha \in \left[-\frac{\pi}{3}, 0\right) \Rightarrow \cos(\alpha) \geq \cos\left(\alpha + \frac{2\pi}{3}\right) \geq \cos\left(\alpha - \frac{2\pi}{3}\right)$$

$$\alpha \in \left(0, \frac{\pi}{3}\right) \Rightarrow \cos(\alpha) \geq \cos\left(\alpha - \frac{2\pi}{3}\right) \geq \cos\left(\alpha + \frac{2\pi}{3}\right) \tag{A8}$$

In the following calculation, the upper sign will refer to the negative and the lower sign will refer to the positive interval

$$\begin{aligned} \delta W(\varrho_{SA}, H, \Pi) &= W_D(\varrho_{SA}, H, \Pi) - W(\varrho_S, H) \\ &= \text{Tr}[\varrho_S H] - \min_\Pi \sum_k \text{Tr}[\varrho_S A(U_k^\dagger H U_k \otimes \Pi_k)] - \left[\text{Tr}[\varrho_S H] - \min_U \text{Tr}[\varrho_S U^\dagger H U]\right] \\ &= \min_U \text{Tr}[\varrho_S U^\dagger H U] - \min_\Pi \sum_k \text{Tr}[\varrho_S A(U_k^\dagger H U_k \otimes \Pi_k)] \\ &= \max_\alpha \left\{\frac{1}{3}(\epsilon_0 + \epsilon_1 + \epsilon_2) - \frac{1}{6}(\epsilon_0[1 + \cos(\alpha)]) + \epsilon_1\left(1 + \cos\left(\alpha \pm \frac{2\pi}{3}\right)\right)\right. \\ &\quad + \epsilon_2\left(1 + \cos\left(\alpha \mp \frac{2\pi}{3}\right)\right) + \epsilon_0\left(1 - \cos\left(\alpha \mp \frac{2\pi}{3}\right)\right) \\ &\quad \left. + \epsilon_1\left(1 - \cos\left(\alpha \pm \frac{2\pi}{3}\right)\right) + \epsilon_2(1 - \cos\alpha)\right\} \\ &= \frac{1}{6}(\epsilon_2 - \epsilon_0)\max_\alpha\left(\cos(\alpha) - \cos\left(\alpha \mp \frac{2\pi}{3}\right)\right) \\ &= \frac{\epsilon_2 - \epsilon_0}{2\sqrt{3}}. \end{aligned} \tag{A9}$$

Now, that we computed the maximal daemonic gain for projective measurements, we compare this with the daemonic gain that can be achieved by using the POVM M, consisting of the effects $\frac{2}{3}P_i$, as defined in Equation (A2). In this case, the conditional states are

$$\gamma_{P_0}^S = \frac{2}{9}\left(|0\rangle\langle 0| + \frac{1}{4}|1\rangle\langle 1| + \frac{1}{4}|2\rangle\langle 2|\right),$$
$$\gamma_{P_1}^S = \frac{2}{9}\left(\frac{1}{4}|0\rangle\langle 0| + |1\rangle\langle 1| + \frac{1}{4}|2\rangle\langle 2|\right),$$
$$\gamma_{P_2}^S = \frac{2}{9}\left(\frac{1}{4}|0\rangle\langle 0| + \frac{1}{4}|1\rangle\langle 1| + |2\rangle\langle 2|\right). \tag{A10}$$

Given the conditional states, we can now compute the daemonic gain as

$$\delta W = \epsilon_0 \left(\frac{1}{3} - \frac{2}{3}\right) + \epsilon_1 \left(\frac{1}{3} - \frac{1}{6}\right) + \epsilon_2 \left(\frac{1}{3} - \frac{1}{6}\right)$$
$$= -\frac{1}{3}\epsilon_0 + \frac{1}{6}(\epsilon_1 + \epsilon_2). \tag{A11}$$

Choosing the Hamiltonian $H = |\epsilon_1\rangle\langle\epsilon_1| + |\epsilon_2\rangle\langle\epsilon_2|$ provides an example where the maximal daemonic gain can not be achieved by using projective measurements because

$$\delta W_{\text{proj}} = \frac{1}{2\sqrt{3}} < \delta W_M = \frac{1}{3}. \tag{A12}$$

Appendix B. Non-Optimal Convergence of the See-Saw Algorithm

In the following, we construct a case in which Algorithm 1 will yield a sequence of values for the daemonic ergotropy that does not converge against the maximal daemonic ergotropy. Consider a state ϱ^{SA} on a system S with a Hamiltonian H and a d-dimensional ancilla A, such that the optimal measurements are rank-one projective measurements as long as only d-outcome measurements are considered. Then, there exists an initialisation of Algorithm 1, such that the sequence of daemonic ergotropies generated by the algorithm limits in the maximal daemonic ergotropy for d-outcome measurements. In order to see this, consider a measurement Π that is optimal among d-outcome measurements. For the effects $\{\Pi_1, \ldots, \Pi_d\}$ one finds d optimal unitaries $\{V_1, \ldots, V_d\}$. We now initialise the algorithm for d^2 outcomes in the following way

$$U_i = V_i, \quad i = 1, \ldots, d-1$$
$$U_i = V_d, \quad i = d, \ldots, d^2. \tag{A13}$$

This implies $\tau_d = \tau_{d+1} = \ldots = \tau_{d^2}$, where $\tau_i = \text{Tr}_S(\varrho^{SA} U_i^\dagger H U_i)$. Hence, the objective of step 2 of the algorithm simplifies to

$$\min_M \sum_{i=1}^{d^2} \text{Tr}(\tau_i M_i) = \min_M \left[\sum_{i=1}^{d-1} \text{Tr}(\tau_i M_i) + \text{Tr}\left(\tau_d \sum_{j=d}^{d^2} M_j\right)\right]. \tag{A14}$$

The value of this expression thus depends on d effects $M_1, \ldots, M_{d-1}, \sum_{j=d}^{d^2} M_j$. In this case, the minimum can by assumption only be achieved if the effects are all rank-one. This implies that the first $d-1$ effects are orthogonal rank-one projectors and the remaining effects are rank-one operators on the remaining one-dimensional subspace and sum up to a rank-one projector. Thus, the algorithm again finds a d-outcome rank-one projective measurement that is optimal among d-outcome measurements. The case that was discussed above is of practical relevance, as we have observed in numerical experiments that randomly initialised unitaries may converge against the configuration stated in Equation (A13).

The example discussed in Appendix A meets the requirement that all optimal two-outcome measurements are rank-one projective measurements. The optimal projective measurements are calculated in Appendix A. Any two outcome measurement in two dimensions with rank-two effects can be considered as a mixture of a rank-one projective measurement with white noise. The only case, in which white noise will not decrease the daemonic ergotropy is, if the conditional states γ_i^S [Equation (A8)] are simultaneously diagonalisable by the same diagonalising unitary and with the same ordering of eigenvalues in diagonal form. This is however not the case, since both states are already diagonal but the eigenvalues are not in the same order.

In the same example, the maximum daemonic ergotropy cannot be achieved with d-outcome measurements.

Appendix C. Proof of Theorem 5

In this Appendix we provide a complete proof of the statement made in Theorem 5, which we repeat here again for easiness of reading. For a quantum-classical state, that is a state that can be cast in the form

$$\varrho_{qc}^{SA} = \sum_j \sigma_j^S \otimes |j\rangle\langle j|^A \qquad (A15)$$

with a set of orthonormal vectors $\{|j\rangle^A\}$ and unnormalised states σ_j^S the following theorem holds.

Theorem A1. *For a quantum-classical state ϱ_{qc}^{SA}, the maximum daemonic ergotropy is*

$$\max_M W_D(\varrho^{SA}, H, M) = \sum_j W(\sigma_j^S, H). \qquad (A16)$$

This value is achieved by performing the projective measurements $P_j = |j\rangle\langle j|^A$ on the ancilla A.

Proof. The first claim follows directly from the second claim using Equation (3). Therefore, we prove the second claim by showing that the daemonic gain achieved through any POVM E with effects E_i and an arbitrary number of outcomes N has an upper bound given by the value corresponding to the use of projective measurements. We start by computing the conditional states

$$\gamma_k^S = \text{Tr}_A\left[\varrho^{SA}(I \otimes E_k)\right] = \sum_{j=1}^d \sigma_j^S \langle j| E_k |j\rangle. \qquad (A17)$$

It can be easily seen that post-processing can never increase the daemonic ergotropy. This allows us to assume, without loss of generality, that all effects are rank-one and use Naimark's extension theorem [30] to write

$$\gamma_k^S = \sum_{j=1}^N \sigma_j^S |\langle j|\phi_k\rangle|^2, \qquad (A18)$$

where $(|\phi_k\rangle\langle\phi_k|)_{k=1}^N$ is the Naimark extension of the operators E_k on the extended ancilla space. Then, $(|\phi_k\rangle)_{k=1}^N$ is an orthonormal basis in the extended ancilla space. We also extend $(|j\rangle)_{j=1}^d$, so $(|j\rangle)_{j=1}^N$ is another orthonormal basis in the extended ancilla space and set $\sigma_j^S = 0, \forall j > d$. We can now interpret $|\langle j|\phi_k\rangle|^2$ as entries of a doubly stochastic matrix and apply the Birkhoff-von Neumann theorem [31], which allows us to express this doubly stochastic matrix as a convex combination of permutation matrices $\pi^{(n)} = \left(\pi_{jk}^{(n)}\right)_{jk}$. This yields

$$\gamma_k^S = \sum_{j=1}^N \sigma_j \sum_n p_n \pi_{jk}^{(n)} \qquad (A19)$$

with probabilities p_n.

We insert this result into the formula of the daemonic ergotropy

$$W_D(\varrho^{SA}, H, M) = \text{Tr}(\varrho^S H) - \sum_k \min_{U_k} \text{Tr}(U_k \gamma_k^S U_k^\dagger H). \tag{A20}$$

As we are interested in the optimal measurement, our only concern is the second term

$$\sum_{k=1}^N \min_U \text{Tr}(U \gamma_k^S U^\dagger H)$$

$$= \sum_{k=1}^N \min_U \text{Tr}(U \sum_{j=1}^N \sigma_j \sum_n p_n \pi_{jk}^{(n)} U^\dagger H)$$

$$\geq \sum_{k,j,n} p_n \pi_{jk}^{(n)} \min_U \text{Tr}(U \sigma_j U^\dagger H)$$

$$= \sum_n p_n \sum_j \min_U \text{Tr}(U \sigma_j U^\dagger H) \sum_k \pi_{jk}^{(n)}$$

$$= \sum_j \min_U \text{Tr}(U \sigma_j U^\dagger H), \tag{A21}$$

which is bounded from below by the value that is achieved for the projective measurement $P_j = |j\rangle\langle j|$, as stated above. □

References

1. Goold, J.; Huber, M.; Riera, A.; del Rio, L.; Skrzypczyk, P. The role of quantum information in thermodynamics—A topical review. *J. Phys. A: Math. Theor.* **2016**, *49*, 143001. [CrossRef]
2. Kosloff, R. Quantum Thermodynamics: A Dynamical Viewpoint. *Entropy* **2013**, *15*, 2100–2128. [CrossRef]
3. Vinjanampathy, S.; Anders, J. Quantum Thermodynamics. *Contemp. Phys.* **2015**, *57*, 1. [CrossRef]
4. Gelbwaser-Klimovsky, D.; Niedenzu, W.; Kurizki, G. Thermodynamics of quantum systems under dynamical control. *Adv. Atom. Mol. Opt. Phys.* **2015**, *64*, 329.
5. Scully, M.; Zubairy, M.S.; Agarwal, G.S.; Walther, H. Extracting work from a single heat bath via vanishing quantum coherence. *Science* **2003**, *299*, 862. [CrossRef] [PubMed]
6. Karimi B.; Pekola, J. P. Otto refrigerator based on a superconducting qubit: Classical and quantum performance. *Phys. Rev. B* **2016**, *94*, 184503. [CrossRef]
7. Uzdin, R.; Levy, A.; Kosloff, R. Equivalence of Quantum Heat Machines, and Quantum-Thermodynamic Signatures. *Phys. Rev. X* **2015**, *5*, 031044. [CrossRef]
8. Elouard, C.; Jordan, A. N. Efficient Quantum Measurement Engines. *Phys. Rev. Lett.* **2018**, *120*, 260601. [CrossRef] [PubMed]
9. Elouard, C.; Herrera-Martí, D.; Huard, B.; Auffèves, A. Extracting Work from Quantum Measurement in Maxwell's Demon Engines. *Phys. Rev. Lett.* **2017**, *118*, 260603. [CrossRef]
10. Yi, J.; Talkner, P.; Kim, Y. W. Single-temperature quantum engine without feedback control. *Phys. Rev. E* **2017**, *96*, 022108. [CrossRef]
11. Gelbwaser-Klimovsky, D.; Erez, N.; Alicki, R.; Kurizki, G. Work extraction via quantum nondemolition measurements of qubits in cavities: Non-Markovian effects. *Phys. Rev. A* **2013**, *88*, 022112. [CrossRef]
12. Jacobs, K. Second law of thermodynamics and quantum feedback control: Maxwell's demon with weak measurements. *Phys. Rev. A* **2009**, *80*, 012322. [CrossRef]
13. Abah, O.; Paternostro, M. Implications of non-Markovian dynamics on information-driven engine. *arXiv* **2019**, arXiv:1902.06153.
14. Hovhannisyan, K.; Perarnau-Llobet, M.; Huber, MM.; Acín, A. Entanglement Generation is Not Necessary for Optimal Work Extraction. *Phys. Rev. Lett.* **2013**, *111*, 240401. [CrossRef] [PubMed]
15. Perarnau-Llobet, M.; Hovhannisyan, K.; Huber, M.; Skrzypczyk, P.; Brunner, N.; Acín, A. Extractable Work from Correlations. *Phys. Rev. X* **2015**, *5*, 041011. [CrossRef]

16. Fusco, L.; Paternostro, M.; De Chiara, G. Work extraction and energy storage in the Dicke model. *Phys. Rev. E* **2016**, *94*, 052122. [CrossRef] [PubMed]
17. Campisi, M.; Fazio, R. Dissipation, Correlation and Lags in Heat Engines. *J. Phys. A: Math. Theor.* **2016**, *49*, 345002. [CrossRef]
18. Francica, G.; Goold, J.; Plastina, F.; Paternostro, M. Daemonic Ergotropy: Enhanced Work Extraction from Quantum Correlations. *NPJ Quant. Inf.* **2018**, *3*, 12. [CrossRef]
19. Allahverdyan, A.; Balian, R.; Nieuwenhuizen, T. Maximal work extraction from finite quantum systems. *Europhys. Lett.* **2004**, *67*, 565. [CrossRef]
20. Mirsky, L. A Trace Inequality of John von Neumann. *Monatshefte für Mathematik* **1975**, *79*, 303–306. [CrossRef]
21. Heinosaari, T.; Ziman, M. *The Mathematical Language of Quantum Theory: From Uncertainty to Entanglement*; Cambridge University Press: Cambridge, UK, 2011.
22. Chiribella, G.; D'Ariano, G.; Schlingemann, D. How Continuous Quantum Measurements in Finite Dimensions Are Actually Discrete. *Phys. Rev. Lett.* **2007**, *98*, 190403. [CrossRef] [PubMed]
23. Uola, R. (Group of Applied Physics, University of Geneva, Genève, Switzerland). Private communication, 2018.
24. Olliver, H.; Zurek, W. Quantum Discord: A Measure of the Quantumness of Correlations. *Phys. Rev. Lett.* **2001**, *88*, 017901. [CrossRef] [PubMed]
25. Henderson, L.; Vedral, V. Classical, quantum and total correlations. *J. Phys. A: Math. Gen.* **2001**, *34*, 6899. [CrossRef]
26. Modi, K.; Broduteh, A.; Cable, H.; Paterek, T.; Vedral, V. The classical-quantum boundary for correlations: Discord and related measures. *Rev. Mod. Phys.* **2012**, *84*, 1655. [CrossRef]
27. Navascues, M. Research Lines that Lead Nowhere (1): Quantum Discord. Available online: http://schroedingersrat.blogspot.com/2012/10/research-lines-that-lead-nowhere-i.html (accessed on 6 August 2019).
28. Peres, A. Higher order Schmidt decompositions. *Phys. Lett. A* **1995**, *202*, 16. [CrossRef]
29. Neven, A.; Martin, J.; Bastin, T. Entanglement robustness against particle loss in multiqubit systems. *Phys. Rev. A* **2018**, *98*, 062335. [CrossRef]
30. Nielsen, M.; Chuang, I. *Quantum Computation and Quantum Information*; Cambridge University Press: Cambridge, UK, 2000.
31. Lenard, A. Thermodynamical proof of the Gibbs formula for elementary quantum systems. *J. Stat. Phys.* **1978**, *19*, 575. [CrossRef]

© 2019 by the authors. Licensee MDPI, Basel, Switzerland. This article is an open access article distributed under the terms and conditions of the Creative Commons Attribution (CC BY) license (http://creativecommons.org/licenses/by/4.0/).

Article

Thermalization of Finite Many-Body Systems by a Collision Model

Onat Arısoy [1,2,3,*], Steve Campbell [4] and Özgür E. Müstecaplıoğlu [1]

1. Department of Physics, Koç University, İstanbul, Sarıyer 34450, Turkey
2. Institute for Physical Science and Technology, University of Maryland, College Park, MD 20742, USA
3. Chemical Physics Program, University of Maryland, College Park, MD 20742, USA
4. School of Physics, University College Dublin, Belfield, Dublin 4, Ireland
* Correspondence: oarisoy14@ku.edu.tr

Received: 29 October 2019; Accepted: 28 November 2019; Published: 30 November 2019

Abstract: We construct a collision model description of the thermalization of a finite many-body system by using careful derivation of the corresponding Lindblad-type master equation in the weak coupling regime. Using the example of a two-level target system, we show that collision model thermalization is crucially dependent on the various relevant system and bath timescales and on ensuring that the environment is composed of ancillae which are resonant with the system transition frequencies. Using this, we extend our analysis to show that our collision model can lead to thermalization for certain classes of many-body systems. We establish that for classically correlated systems our approach is effective, while we also highlight its shortcomings, in particular with regards to reaching entangled thermal states.

Keywords: collision model; thermalization; many-body quantum systems

---

1. Introduction

Computer simulations of finite many-body systems have been challenging and expanding predictions of statistical mechanics since their first application to test equilibration of an anharmonic crystal modeled by a chain of masses with fixed-ends [1]. While standard methods to investigate equilibration and thermalization of quantum systems are based upon master equations [2], so called quantum collision models are introduced as versatile computational tools for simulating and studying open quantum systems [3,4]. The simplest collision model consists of a two-level system undergoing repeated collisions with environment, or ancilla, two-level systems. It is equivalent to a discrete time Markovian master equation in Lindblad form for the dynamics of the system, for short collision times [5]. Here, we address the question of how to generalize the collision models to finite quantum many-body systems for illuminating their thermalization dynamics.

Intuitively, it is reasonable to obtain a Markovian dynamics of the system using collisions if the colliding ancillae do not interact with any other degrees of freedom since such short time interactions should not allow any significant memory effects. However, the often implicit assumption of stronger interaction than the system Hamiltonian and the neglecting of the bath Hamiltonian are not always valid. Furthermore, using the typical formalism, e.g., [5,6] where energy preserving exchange interactions are considered, results in a dynamics which drives the system to the same state as ancillae, meaning that the result is independent from the system Hamiltonian and homogenization, rather than thermalization, is achieved [7,8]. This problem persists and is compounded for the generalization of collision models for many-body systems [8,9]. Interestingly, [10] derives a Lindblad type master equation for collisions with arbitrary interaction strengths and collision times and establishes that the thermal state of a system at the environment temperature with respect to the Hamiltonian $\hat{H}_0$

is an equilibrium state if $[\hat{U}, \hat{H}_0 + \hat{H}_{bath}]$ where $\hat{U}$ is the unitary evolution operator under the total Hamiltonian. However, setting $\hat{H}_0 = \hat{H}_{system}$ and finding the necessary interaction type and strength to satisfy this commutation property remains as a challenging open problem so far. At variance with this and other works that study collision models starting from a "global" unitary picture [9,10], in this work we propose a master equation derivation inspired by the well-known derivation for a time-independent system–bath interaction in the weak coupling regime [2].

Despite its drawbacks in describing Markovian open system dynamics, quantum collision models are still a good candidate for understanding the quantum thermodynamical phenomena from a microscopic perspective [11]. For example, the microscopic Markovian master equation derivation in [2] does not account for the information loss of the system about its initial state, while it is evident using the collision model that the lost information is kept by the entanglement between the system and ancillae [12]. Another study analyzes the entropy generation and distribution in a collision model and proves the asymptotic factorization of the total density matrix of system and environment into the density matrices of the system and the environment for a two level system in the strong coupling regime [13]. More complex collision models involving ancilla–ancilla collisions allow for the derivation of completely positive non-Markovian dynamics [14–16]. The controllable degree of non-Markovianity and its effect on the dynamics of quantum coherence has been examined [17]. Further attempts to study non-Markovian dynamics are made by using initially entangled ancillae [18], introducing time overlap of two consecutive collisions [6,19,20] and using a two-spin system in which only one of the pair interacts with the environment resulting in a Markovian dynamics for the composite system, while tracing out the spin interacting with the bath gives a non-Markovian dynamics for the remaining spin [21]. The versatility of collision models has resulted in other interesting research directions, such as the introduction of collisions with non-thermalized ancillae to study non-equilibrium effects in quantum thermodynamics [8,11,22–25] and the generation of multi-qubit entanglement via a shuttle qubit colliding with disjoint qubit registers [26].

This work aims to examine the conditions required for thermalization in a Markovian collision model using two level ancillae. To this end, we will first carefully examine the microscopic derivation of a Lindblad master equation for a two level system in the weak coupling regime from [2] and introduce a time dependent interaction Hamiltonian in Section 2, where we also assess each assumption made for the derivation and examine their validity. Section 3 examines our collision model for many-body systems for both non-entangled and entangled energy eigenstates with an example for each of these cases illustrating how our proposed collisional route to many-body thermalization works. Finally, we conclude in Section 4.

2. Derivation and Validity of Lindblad Master Equation

We begin by following the microscopic derivation of the Lindblad master equation given in [2], however allowing for a time-dependent interaction Hamiltonian instead of using the second order approximation of the unitary evolution operator for the system and the ancilla with respect to the collision time [21]. The dynamics of the system and the bath is governed by the Liouville-von Neumann equation

$$\frac{d}{dt}\rho(t) = -i[\hat{H}_I(t), \rho(t)]. \tag{1}$$

Integrating this equation with respect to time and plugging in the expression for $\rho(t)$ in the commutator twice with the assumption of $\text{Tr}_B([\hat{H}_I(t), \rho(0)]) = 0$ we arrive at

$$\frac{d}{dt}\rho(t) = -\int_0^t ds [\hat{H}_I(t), [\hat{H}_I(s), \rho(s)]]. \tag{2}$$

Applying the Born approximation by neglecting system–bath entanglement and the effect of the system on the bath allows us to write an equation for the dynamics of the system by tracing over the bath degrees of freedom

$$\frac{d}{dt}\rho_s(t) = -\int_0^t ds \, \mathrm{Tr}_B([\hat{H}_I(t),[\hat{H}_I(s),\rho_S(s)\otimes\rho_B]]). \tag{3}$$

At this point, the dynamics of the system is still, in general, non-Markovian and we have not made any explicit assumptions about the nature of the interaction. However, the finite time of a given collision may serve to justify the constancy of the bath density matrix along with the weak interaction assumption. Putting aside the validity of Born approximation, we need to explicitly assume that the density matrix of the system does not change significantly during the interaction with a single ancilla, which is justifiable for short collision times, in order to replace the past states of the system with its present state and to obtain the Redfield equation

$$\frac{d}{dt}\rho_s(t) = -\int_0^t ds \, \mathrm{Tr}_B([\hat{H}_I(t),[\hat{H}_I(t-s),\rho_S(t)\otimes\rho_B]]). \tag{4}$$

The standard master equation derivation in [2] for a time-independent interaction Hamiltonian continues with the assumption that the integrand above vanishes quickly enough to extend the integral to infinity with negligible difference on the system dynamics. In our case of short time collisions starting after $t = 0$, this extension is not an assumption to be checked as the integrand is explicitly zeroed out for $s > t$ by the time-dependent strength of the interaction Hamiltonian. For simplicity, we assume that each ancilla interacts with the system once and these collisions start with a period of τ_p and a duration of τ_c.

After explicitly defining our collision model, we can investigate the effects of the finite time interactions on the dynamics. As in the derivation in [2], we will introduce the interaction Hamiltonian in the Schrödinger picture

$$\hat{H}_I = \sum_\alpha \hat{A}_\alpha \otimes \hat{B}_\alpha \tag{5}$$

where the Hermitian operators $\hat{A}_\alpha$ and $\hat{B}_\alpha$ act on the system and the bath respectively. After decomposing the operators $\hat{A}_\alpha$ into operators $\hat{A}_\alpha(\omega)$ based on the energy transitions with frequency ω generated on the eigenstates of the system Hamiltonian and plugging the interaction picture interaction Hamiltonian in Equation (4), we obtain

$$\frac{d}{dt}\rho_s(t) = \sum_{\omega,\omega'}\sum_{\alpha,\beta} e^{it(\omega'-\omega)}\Gamma_{\alpha\beta}(\omega)\left(\hat{A}_\beta(\omega)\rho_s(t)\hat{A}_\alpha^\dagger(\omega')\right.$$
$$\left. - \hat{A}_\alpha^\dagger(\omega')\hat{A}_\beta(\omega)\rho_s(t)\right) + \text{h.c.} \tag{6}$$

where $\Gamma_{\alpha\beta}(\omega)$ is the one-sided Fourier transform of the reservoir correlation functions

$$\Gamma_{\alpha\beta}(\omega) = \int_0^\infty ds\, e^{is\omega}\,\mathrm{Tr}_B(\hat{B}_\alpha^\dagger(t)\hat{B}_\beta(t-s)) \tag{7}$$

where operators are defined in the interaction picture.

For the evaluation of bath correlation spectra, we must specify our open system setup which consists of the same basic ingredients as [6,19,20,25]. We first consider a two-level system with time-independent Hamiltonian

$$\hat{H}_S = h_s\hat{\sigma}_z. \tag{8}$$

The reservoir consists of, an in principle infinite number of, two-level systems prepared at an inverse temperature $\beta_b = 1/(k_B T)$ for a bath Hamiltonian

$$\hat{H}_B = \sum_{n=1}^{N} h_b \hat{\sigma}_{zn}. \tag{9}$$

where the index n indicates that the operator acts on the n-th spin of the reservoir. The time-dependent interaction Hamiltonian in the Schrödinger picture is given by

$$\hat{H}_I = \sum_{n=1}^{N} g_n(t) \hat{\sigma}_x \otimes \hat{\sigma}_{xn} \tag{10}$$

where the operator without index acts on the system. For simplicity, we assume that the interaction strength is exactly zero before and after the interaction, remains constant during the collision, and has the same magnitude for all collisions. It should be noted that the interaction in Equation (10) is different from the often considered partial swap case which is known to lead to homogenization [4] rather than thermalization [7]. Knowing the collision period and duration, we can now define the time-dependent interaction strengths as

$$g_n(t) = \theta(t - (n-1)\tau_p)\theta((n-1)\tau_p + \tau_c - t)g \tag{11}$$

where θ denotes the Heaviside step function.

Before explicitly calculating the bath correlation spectra, we can make some simplifications. As each ancilla has one interaction component in the form of Equation (5), the indices α and β in fact denote the index of the corresponding ancilla. Also, knowing that all ancillae are prepared in a thermal state, it is easy to prove that cross correlations vanish and we can arrange Equation (6) in the form

$$\frac{d}{dt}\rho_s(t) = \sum_{\omega,\omega'} \sum_{n=1}^{N} e^{it(\omega'-\omega)} \Gamma_n(\omega) \left(\hat{A}_n(\omega) \rho_s(t) \hat{A}_n^\dagger(\omega') - \hat{A}_n^\dagger(\omega') \hat{A}_n(\omega) \rho_s(t) \right) + \text{h.c.} \tag{12}$$

After some manipulation, we find the explicit form of reservoir correlation spectra

$$\Gamma_n(\omega) = g^2 \theta(t - (n-1)\tau_p)\theta((n-1)\tau_p + \tau_c - t)$$
$$\int_0^\infty ds\, e^{is\omega} (\rho_{ee}^n e^{2ih_b s} + \rho_{gg}^n e^{-2ih_b s}) \theta((t-s) - (n-1)\tau_p)\theta((n-1)\tau_p + \tau_c - (t-s))$$
$$= g^2 \theta(t - (n-1)\tau_p)\theta((n-1)\tau_p + \tau_c - t) \int_0^{t-(n-1)\tau_p} ds\, e^{is\omega}(\rho_{ee}^n e^{2ih_b s} + \rho_{gg}^n e^{-2ih_b s}) \tag{13}$$

where ρ_{ee}^n and ρ_{gg}^n are excited and ground populations of n-th ancilla. It is clear that the bath correlation spectra are time-dependent and they are zeroed out by the step functions before or after the collision. We must evaluate this expression for the cases $\omega = \pm 2h_b$ and $\omega \neq \pm 2h_b$ separately,

$$\Gamma_n(\omega \neq \pm 2h_b, t) = -ig^2 \left(\frac{\rho_{ee}^n(\exp(i(t - (n-1)\tau_p)(\omega + 2h_b)) - 1)}{\omega + 2h_b} + \frac{\rho_{gg}^n(\exp(i(t - (n-1)\tau_p)(\omega - 2h_b)) - 1)}{\omega - 2h_b} \right) \tag{14}$$

If $\omega = \pm 2h_b$, one of the complex exponentials in the integrand simplifies and gives a linearly growing term

$$\Gamma_n(\omega = -2h_b, t) = g^2(\rho_{ee}^n(t - (n-1)\tau_p)) - \frac{i\rho_{gg}^n(\exp(i(t - (n-1)\tau_p)(\omega - 2h_b)) - 1)}{\omega - 2h_b}$$
$$\Gamma_n(\omega = 2h_b, t) = g^2(\rho_{gg}^n(t - (n-1)\tau_p)) - \frac{i\rho_{ee}^n(\exp(i(t - (n-1)\tau_p)(\omega + 2h_b)) - 1)}{\omega + 2h_b} \tag{15}$$

The final step of the derivation of a Lindblad type master equation is the decomposition of bath correlation spectra into its real and imaginary parts. The imaginary part results in an additional Hamiltonian term, the Lamb shift acting on the system. However, as this is not relevant to the equilibration of the system, we will neglect it in what follows. As we explicitly show in Figure 1 it is also reasonable to neglect situations where ancillae spins are not on resonance with the system, i.e., we only consider $h_b = h_s$. In this case, the bath correlation spectra consists of a real and linearly growing term and a rotating term with real and complex parts. The linearly growing term generates a dynamics similar to a Lindblad master equation with time-independent interactions, while the real part of the rotating term can be neglected assuming that the relaxation of the system is much slower than the dynamics of the closed system. The master equation in Lindblad form can be obtained after applying these assumptions to Equation (12) together with the secular approximation resulting in

$$\frac{d}{dt}\rho_s(t) = \mathrm{Re}(\Gamma(2h_s,t))(\hat{\sigma}_-\rho_s(t)\hat{\sigma}_+ - \frac{1}{2}\{\hat{\sigma}_+\hat{\sigma}_-,\rho_s(t)\}) + \mathrm{Re}(\Gamma(-2h_s,t))(\hat{\sigma}_+\rho_s(t)\hat{\sigma}_- - \frac{1}{2}\{\hat{\sigma}_-\hat{\sigma}_+,\rho_s(t)\}) \quad (16)$$

where the Γ function contains the information about all of the collisions

$$\mathrm{Re}(\Gamma(\omega,t)) = g^2 \sum_{n=1}^{N} (\delta'(\omega - 2h_b)\rho_{gg}^n + \delta'(\omega + 2h_b)\rho_{ee}^n)(t - (n-1)\tau_p)\theta(t - (n-1)\tau_p)\theta((n-1)\tau_p + \tau_c - t), \quad (17)$$

where N denotes the number of ancilla spins. The function $\delta'(\omega)$ is defined as one for $\omega = 0$ and zero elsewhere, not to be confused with Dirac delta function. Note that this equation neglects the case where the ancilla is not in resonance with the system and it is used throughout Section 3. However, the off-resonance effects in Figure 1 need to be interpreted using the bath correlation spectrum described in Equation (14).

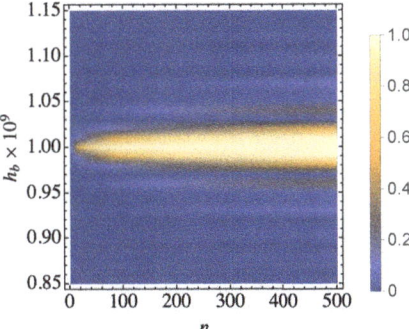

Figure 1. Simulation results for the thermalization of a single two-level system using our collision model. We show the fidelity of the system with the target thermal state as a function of the bath ancilla frequency and number of collisions. We clearly see that thermalization occurs when the system interacts with bath frequencies that are on resonance. We have fixed $T = 10$ mK, $g = 1$ MHz, $h_s = 1$ GHz, $t = 200$ ns, and $\rho_s(0) = |1\rangle\langle 1|$.

The transition from Equation (12) to Equation (16) takes the secular approximation for granted, however it can be justified by some assumptions relating three different time scales of the open system dynamics: The natural evolution times of the system and ancillae and the duration of the collision, all of which play a critical role in constraining the validity of the derived master equation. We assume that the interaction vanishes before any significant change on the density matrix of the ancilla can happen. We also assume that the variation of the system state during one collision is small, which further constrains the maximum collision time. On the other hand, we also want to eliminate the rotating terms of the bath correlation spectra by averaging them over multiple periods of the system dynamics with a slow relaxation of the system which leads to a lower bound of the collision duration.

After justifying the derivation of Equation (16), it is straightforward to find the Kubo–Martin–Schwinger (KMS) condition for n-th collision exploiting the fact that the ancillae are prepared in a thermal state, resulting in vanishing cross bath correlations.

$$\frac{\text{Re}(\Gamma_n(2h_s,t))}{\text{Re}(\Gamma_n(-2h_s,t))} = \exp(2\beta h_s) = \frac{\rho_{gg}^n}{\rho_{ee}^n} = \exp(2\beta_b h_s) \qquad (18)$$

The interpretation of this equation is obvious: The thermal state of the system at the inverse bath temperature β_b is the unique steady state of the Markovian dynamics generated by collisions with ancillae prepared in thermal state [2]. This result was also predicted in complementary works on collision models [5,25] derived using different parameter regimes.

In Figure 1 we simulate our collision model sweeping through a range of frequencies for the bath ancillae and show the final state fidelity between the system and its target thermal state. The simulation consists of the unitary evolution of the system and ancillae during the collision time with the sum of system, bath, and interaction Hamiltonians described above and the ancillae are traced out after each collision without interacting again with the system or other with ancillae. We clearly see that when the ancillae are close to resonance the collision model leads to thermalization of the system. Conversely, when the ancillae are far detuned from h_s we find the system dynamics are almost frozen. This result can be predicted theoretically by calculating the real part of bath correlation spectrum without assuming resonance. Equation (14) has two terms which are inversely proportional to the difference between the transition frequency ω and $\pm 2h_b$. Assuming a small detuning from either $2h_b$ or $-2h_b$, the other term becomes negligibly small. After dropping the small term, evaluating the real part for the other part gives

$$\text{Re}(\Gamma_n(\mp 2h_s, t)) = \frac{\rho_{ee,gg}^n g^2 \sin(\delta t)}{\delta}, \delta = \mp 2h_s \pm 2h_b \qquad (19)$$

ignoring the Heaviside step functions and taking the beginning of each collision as $t = 0$. Its limit for $\delta \to 0$ recovers the case of resonance. The off-resonance dynamics depend heavily on the product $\delta \tau_c$. As the average of sine function over a period is zero, we can conclude that the effect of the dissipative term should be negligible if the product $\delta \tau_c = 2k\pi$ where k is an integer and the dynamics is slow enough. On the other hand, in the case where the product is an odd multiple of π, the average of sine function is not zeroed out and we observe thermalization as seen in Figure 1. Furthermore, it is straightforward to prove that the fastest thermalization is achieved in the case of resonance using the identity

$$\frac{\sin(\delta t)}{\delta} \geq t, t \geq 0. \qquad (20)$$

The results in Figure 1 confirm the range of validity of our master equation and are in keeping with other results in the literature [25]. Furthermore, the clear importance of on-resonance ancillae indicates that, under suitable constraints, only particular bath frequencies are relevant for ensuring the system thermalizes. Thus, we can exploit this feature to explore the requirements for achieving thermalization for many-body systems.

3. Thermalization of Finite Many-Body Systems

3.1. Classically Correlated Systems

Let us consider the 1D Ising chain described by the Hamiltonian

$$\hat{H}_S = \sum_{i=1}^{N} h_i \hat{\sigma}_{zi} + \sum_{i=1}^{N-1} J_i \hat{\sigma}_{zi} \hat{\sigma}_{z(i+1)}. \qquad (21)$$

As stressed in the previous section, to achieve thermalization we require the driving frequency of the system and the ancillae to be the same. In the case of interacting many-body systems, it should be clear that there will be a range of frequencies, each of which will be related to the various transition frequencies of the many-body system. Thus, to examine the requirements to reach thermalization we use the expression of the interaction Hamiltonian in the form

$$\hat{H}_I = \sum_{i=1}^{N} \sum_{n=1}^{N_i} \sum_{\omega} g_{i,n}(t) \hat{\sigma}_{xi}(\omega) \otimes \hat{\sigma}_{x(i,n)} \tag{22}$$

where sum over ω denotes the decomposition of each spin-ancilla collision operator into the different energy transitions it generates. We can make a temporary simplification to make the illustration of many-body system thermalization easier by replacing the ancillae with a set of harmonic oscillators forming a continuous spectrum prepared at an inverse temperature $\beta_b = 1/(k_B T)$. In this case, we can find the energy transitions generated by each term of the interaction Hamiltonian with a partition of the Hilbert space of the whole system based on each nearest neighbor configuration with respect to a reference spin denoted as i. We can write all terms of the system Hamiltonian involving i-th spin as

$$\hat{H}^i = (J_{i-1}\hat{\sigma}_{z(i-1)} + h_i + J_i\hat{\sigma}_{z(i+1)})\hat{\sigma}_{zi} \tag{23}$$

where $i \neq 1, N$ as the first and last spins of the Ising chain do not have a left and right neighbor, respectively. The Hamiltonian at the end points $i = 1, N$ can be found by omitting the term corresponding to the lacking neighbors $i = 0, N+1$ in the above equation.

We can now define the transition frequencies generated by flipping the i-th spin in terms of the state of neighbor spins

$$\begin{aligned}
\omega\left(\left|\uparrow^{i-1}\uparrow^{i+1}\right\rangle\right) &= 2(J_{i-1} + h_i + J_i), \\
\omega\left(\left|\uparrow^{i-1}\downarrow^{i+1}\right\rangle\right) &= 2(J_{i-1} + h_i - J_i), \\
\omega\left(\left|\downarrow^{i-1}\uparrow^{i+1}\right\rangle\right) &= 2(-J_{i-1} + h_i + J_i), \\
\omega\left(\left|\downarrow^{i-1}\downarrow^{i+1}\right\rangle\right) &= 2(-J_{i-1} + h_i - J_i).
\end{aligned} \tag{24}$$

Decomposing the operator $\hat{\sigma}_x$ as

$$\hat{\sigma}_{xi} = \hat{\sigma}_{-i} + \hat{\sigma}_{+i}, \tag{25}$$

we obtain two dissipators for each term of the interaction Hamiltonian. The frequencies in Equation (24) correspond to the transitions generated by $\hat{\sigma}_{-i}$ while their negatives correspond to $\hat{\sigma}_{+i}$. Expressing the frequencies as a function of nearest neighbor configuration for each spin results in the master equation

$$\frac{d}{dt}\rho_s = \sum_{i=1}^{N} \sum_{\{s_i\}} \left(\gamma_i(\omega(s_i)) D(\rho_s, \hat{\sigma}_{-i}^{s_i}) + \gamma_i(-\omega(s_i)) D(\rho_s, \hat{\sigma}_{+i}^{s_i}) \right). \tag{26}$$

Here, $\{s_i\}$ is a short hand notation for the respective arguments of the frequencies in Equation (24), corresponding to the set of basis vectors of the Hilbert space of the neighbor spins of i-th spin. The notation $\hat{\sigma}_{\pm i}^{s_i}$ implies that this operator can be decomposed as

$$\begin{aligned}
\hat{\sigma}_{-i}^{s_i} &= |\downarrow\rangle_i \langle\uparrow|_i \otimes |s_i\rangle \langle s_i| \\
\hat{\sigma}_{+i}^{s_i} &= (\hat{\sigma}_{-i}^{s_i})^\dagger
\end{aligned} \tag{27}$$

and $D(\rho, \hat{o})$ is defined by

$$D(\rho, \hat{o}) = \hat{o}\rho\hat{o}^\dagger - \frac{1}{2}\{\hat{o}^\dagger\hat{o}, \rho\}. \tag{28}$$

Although Equation (26) is derived for a continuous set of harmonic oscillators, each term appearing in the double sum is similar to the master equation for a two-level system with the driving frequency depending on the nearest neighbor configuration. Therefore, the implementation of a similar master equation with collisions generating one spin flip operations with ancillae driven at the frequencies of single spin transitions, as illustrated in Figure 2, is possible if the secular approximation is valid such that the ancillae cannot generate any transitions other than those corresponding to its driving frequency. The results of Section 2 on the KMS conditions for the bath correlation spectra can be generalized for the master equation of 1D Ising model and this ensures that if all ancillae are prepared at an inverse temperature β_b, the thermal state of the system at the same temperature is a steady state of the master equation [2]. However, the uniqueness of the stationary solution requires additional constraints. A sufficient condition for the uniqueness can be stated as follows [27,28]:

Condition 1. *Let L be the Lindblad superoperator describing the time derivative of the density matrix and $\hat{\sigma}_{\pm i}(\omega(s_i))$ operators the generators of L. The dynamical semigroup generated by L is relaxing in the sense that it drives the density matrix to a unique final state as time tends to infinity regardless of the initial state if the linear span of the generators is an adjoint set and the bicommutant of the generators is the set of all bounded operators acting on the Hilbert space of the system.*

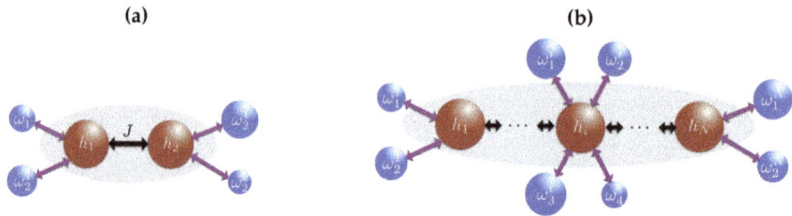

Figure 2. Sketch of our proposed collision model thermalizing (**a**) a two-spin Ising model and (**b**) a many-body spin model. Complete thermalization requires separate ancillae each corresponding to a spin flip transition frequency of the system.

In order to check the applicability of Condition 1 to the thermal bath with local system–bath interactions, we start by checking the adjoint property of the linear span of generators. As established in [27], this follows since $\hat{\sigma}_{+i}(\omega(s_i)) = \hat{\sigma}^{\dagger}_{-i}(\omega(s_i))$, meaning that the adjoint of each generator is also a generator. The second property is easy to prove using the fact that $\hat{\sigma}_{\pm}$ operators only commute with themselves and the identity operator and the only operator commuting with all $\hat{\sigma}_{\pm i}(\omega(s_i))$ for all i and s_i is the identity operator.

To simulate thermalization for a two-site Ising model, Equation (21) with $N = 2$, we require collisions corresponding to the one-spin flip transition frequencies as illustrated in Figure 2a. As each of the spin has a single neighbor, there are two nearest neighbor configurations, resulting in a total of four energy transitions for the whole system. For larger systems, each spin in the bulk of the chain has four different energy transitions and requires more ancillae to successfully thermalize, as shown in Figure 2b.

We implement our collision model for the two-site Ising chain, considering when the collisions with the various ancillae happen "sequentially", i.e., the whole system collides with one of the ancillae corresponding to one of the energy transitions at a time and the colliding ancilla is subsequently traced out before the next collision occurs. We also consider "simultaneously" occurring collisions where the whole system interacts with all of the four ancillae corresponding to different energy transitions at once, after which they are traced out. The minimum energy states are up-down and down-up states and these states cannot be prepared by a local master equation as the collisions are identical, verifying the effect of the system Hamiltonian on open system dynamics resulting in a global master equation. In Figure 3 we show that our collision model gives rise to thermalization for interacting systems. Furthermore, as the cross bath correlations vanish for a thermal bath, we expect that a time overlap between the collisions (such as that which occurs in the simultaneous collision case) does not change the form of the equation, and our numerical results confirm that both approaches generate an almost identical evolution.

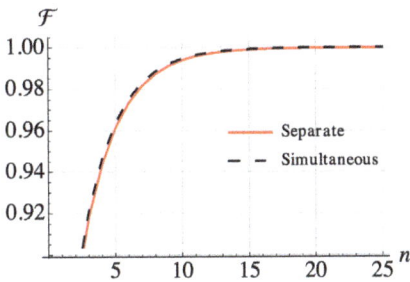

Figure 3. Simulation of a 2-spin Ising model with parameters $h_1 = h_2 = 500$ MHz and $J = 1$ GHz and corresponding transition frequencies $\omega_1 = \omega_3 = 1.5$ GHz and $\omega_2 = \omega_4 = 0.5$ GHz. All ancilla-system spin coupling strengths are set as 1 MHz and the collision times are fixed as 400 ns. Fidelity after each step consisting of one collision for each one spin transition frequency with respect to the thermal state of the system at the temperature of ancilla spins $T_b = 10$ mK. The initial state is the thermal state of the system at infinite temperature.

3.2. Entangled Systems

The Ising model considered in the previous analysis has eigenvectors which are product states without any entanglement among the spin sites. In this section we elaborate on the validity of our collision model for realizing thermalization in more generic many-body systems, particularly those that exhibit entanglement. Addressing such an issue in full generality is a formidable task. Indeed, unlike in the case of non-entangled eigenstates where the generation of single-spin transitions for each interacting neighbor configuration was sufficient, even determining the minimum necessary number of collisions for the uniqueness of the equilibrium state is difficult for entangled states. As such we will restrict to a specific example in this section.

We begin our discussion by reminding that the matrix representation of any Hamiltonian has an eigenvalue decomposition in the form

$$H_s = UDU^\dagger \tag{29}$$

where D is a diagonal matrix with the values of eigenenergies on the diagonal, U is a unitary matrix such that its columns are the eigenstates of the Hamiltonian. Following our master equation derivation, each term of the interaction Hamiltonian is decomposed into different energy transitions, giving rise to operators in the form

$$\hat{A}_{kl} = |\psi'_k\rangle \langle \psi'_l| \tag{30}$$

where $|\psi'_k\rangle$ denotes the k-th eigenstate of the Hamiltonian, which is also denoted by the k-th column of the matrix U. This simple form of the energy transition operators can also be expressed in the basis consisting of the Kronecker product of the bases of the subsystems as

$$\hat{A}_{kl} = \sum_{i=1}^{N} \sum_{j=1}^{N} a_{ij,kl} |i\rangle \langle j| \tag{31}$$

where N is the dimension of the Hilbert space of the system and states i and j are selected from the basis constructed by the Kronecker product of the subsystems, therefore these states are not entangled. Knowing that the k-th column of the matrix U is equal to $|\psi'_k\rangle$, we can write

$$a_{ij,kl} = U^*_{ki} U_{lj} \tag{32}$$

where U_{ab} denotes the element of U at the a-th row and b-th column.

The existence of coefficients $a_{ij,kl}$ indicates that there is a one-to-one linear map from the vectors in the basis of eigenstates to the vectors in the Kronecker product basis. Furthermore, we can vectorize the indices i and j into one index u and the indices k and l into another index v. By these reductions, we can express our linear map in the form of a matrix M such that

$$M_{uv} A_v = x_u \tag{33}$$

where the vectors A and x are the vectorized representations of an operator in the basis of eigenstates and in the Kronecker product basis, respectively, with the sum running over the repeated index v. As the matrix M represents a one-to-one linear map, its inverse exists and any vector A_u can be expressed in the form

$$M^{-1}_{uv} x_v = A_u. \tag{34}$$

Using this expression, we can conclude that a single subsystem transition generated by the interaction Hamiltonian, which can be expressed with a vector x_v, with one non-zero element can generate multiple energy transitions by the multiplication by the inverse of the matrix M used for conversion into eigenstate basis. In this case, the thermalization conditions depend on the structure of the matrix M, however the thermalization of any many-body system is in principle possible with a sufficient number of energy transitions generated by the collisions with two-level ancillae driven at the corresponding transition frequencies, the appropriate choice of interaction Hamiltonian, and the validity of our assumptions for the master equation.

As a concrete example consider a two-spin anisotropic XY-model with Dzyaloshinskii–Moriya (DM) interaction in z-direction

$$\hat{H}_{XY} = J(\hat{\sigma}_{x1}\hat{\sigma}_{x2} - \hat{\sigma}_{y1}\hat{\sigma}_{y2} + \hat{\sigma}_{x1}\hat{\sigma}_{y2} - \hat{\sigma}_{y1}\hat{\sigma}_{x2}) \tag{35}$$

with the eigenstates and eigenenergies [29]

$$|\psi_{1,2}\rangle = \frac{|\downarrow\downarrow\rangle \pm |\uparrow\uparrow\rangle}{\sqrt{2}}, \quad |\psi_{3,4}\rangle = \frac{|\downarrow\uparrow\rangle \mp i|\uparrow\downarrow\rangle}{\sqrt{2}};$$
$$E_{1,3} = 2J, \quad E_{2,4} = -2J. \tag{36}$$

Using the definition of operators $\hat{A}_{kl}$, we can express the one spin flip operators as

$$\begin{aligned}
|\downarrow\downarrow\rangle\langle\uparrow\downarrow| &= -i(\hat{A}_{13} + \hat{A}_{23} - \hat{A}_{14} - \hat{A}_{24})/2 \\
|\uparrow\uparrow\rangle\langle\uparrow\downarrow| &= i(\hat{A}_{13} - \hat{A}_{23} + \hat{A}_{14} - \hat{A}_{24})/2 \\
|\downarrow\downarrow\rangle\langle\downarrow\uparrow| &= (\hat{A}_{13} + \hat{A}_{23} + \hat{A}_{14} + \hat{A}_{24})/2 \\
|\uparrow\uparrow\rangle\langle\downarrow\uparrow| &= (\hat{A}_{13} - \hat{A}_{23} + \hat{A}_{14} - \hat{A}_{24})/2.
\end{aligned} \tag{37}$$

We can then describe the spin ladder operators acting on the first site as

$$\begin{aligned}
\hat{\sigma}_{-1} &= |\downarrow\uparrow\rangle\langle\uparrow\uparrow| + |\downarrow\downarrow\rangle\langle\uparrow\downarrow| \\
&= \frac{1}{2}(\hat{A}_{31} - \hat{A}_{32} + \hat{A}_{41} - \hat{A}_{42} + i(-\hat{A}_{13} - \hat{A}_{23} + \hat{A}_{14} + \hat{A}_{24})) \\
\hat{\sigma}_{+1} &= \hat{\sigma}^\dagger_{-1}.
\end{aligned} \tag{38}$$

It is clear that the $\hat{A}_{41}$ and $\hat{A}_{32}$ terms of the ladder operators and their Hermitians generate state transitions with non-zero energy difference. Other state transitions are between the states with the *same* energy which cannot be generated via with collisions with ancilla spins which do not have internal energy as we have assumed $h_s, h_b \gg g$. If the zero energy transitions were allowed, we could make transitions from any state of the system to another state using intermediate transitions, impling the uniqueness of the thermal state as the equilibrium point of the dynamics [30]. In our case, this condition is not satisfied, and this leads to the equilibrium state of the system exhibiting an initial state dependence. Our numerical simulations in Figure 4 show that if the system is initially prepared in some thermal state, but not in equilibrium with the bath, a Gibbsian thermal state at the environment temperature is achieved. However, it is not guaranteed for generic non-equilibrium initial states, such as $|1\rangle\langle 1|$. We understand this as follows: the choice of initial state as a thermal state at some temperature guarantees that the population of the states having the same energy is equal, thus implying that the zero frequency transition terms will not contribute to the dynamics of the system even if they are generated by the collisions. This means we can assume that the zero frequency terms exist and consequently the equilibrium state is the thermal state at the environment temperature.

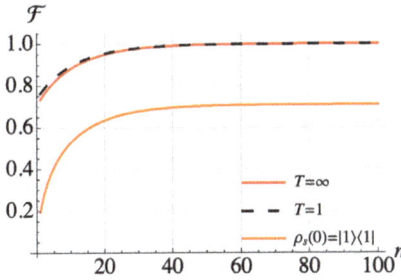

Figure 4. Simulation of a 2-spin anisotropic XY model with $J = 1$ GHz with different initial states of the system. Collision time is set as 400 ns with an ancilla-system interaction strength of 1 MHz. Fidelity after each collision between an ancilla driven with the sole non-zero transition frequency of the system $\omega = 4$ GHz and the first spin of the system each one spin transition frequency with respect to the thermal state of the system at the temperature of ancilla spins $T_b = 10$ mK.

Another possible issue regarding thermalization of entangled many-body systems by our collision model is the additional terms of the master equation due to the non-vanishing bath cross correlations arising due to the decomposition of each term of the interaction Hamiltonian acting on a single subsystem into multiple energy transition terms, which implies that the bath operator of those energy transition terms are the same. For this reason, the positive definiteness of the bath correlation matrix for every frequency needs to be asserted for the uniqueness of the equilibrium state [27].

In summary, our example of two spin anisotropic XY model shows that our collision model can generate multiple energy transitions without the explicit calculation of the M matrix. Although thermalization is not guaranteed, this analysis nevertheless provides insight about how an entangled many-body system with non-degenerate energy levels can be thermalized as long as the secular approximation used in the master equation derivation remains valid and the bath correlation matrix is positive definite.

4. Conclusions

In this work we have presented a collision model using two level ancillae that leads to thermalization in the weak coupling regime, even for certain finite many-body systems. By carefully assessing the relevant timescales present, we showed that when the ancillae are tuned inline with the transition frequencies of the system, thermalization can be achieved. This is at variance with other schemes commonly examined in the literature where system and environment interact via a partial swap [6,20]. Our master equation derivation for 1D Ising model can be straightforwardly generalized to N-dimensional spin lattices by redefining the sums over the Hilbert space of neighbor spins. In the case of Ising spin lattices with more than one dimension, the system can be tuned to be an integrable or non-integrable system depending whether the external magnetic fields are turned off or on respectively [31] and our collision model achieves thermalization in both of the cases. If the eigenstates of the system Hamiltonian are entangled, our collision model gives valuable insight on the dependence of equilibrium state on the initial condition; in particular reveals the conditions to engineer Gibbsian thermal state at the environment temperature. Remarkably, for entangled eigenstates, the decomposition of single-spin transition operators into multiple energy transition operators may remove the necessity of bath interaction with each spin in the system.

Beyond the clear interest in understanding the phenomenology of thermalization using a collision model and its possible extensions to non-Markovian and non-equilibrium dynamics, our collision model also can be viewed as a versatile and implementable artificial environment acting as a temperature knob, as similarly considered in [30,32]. Contrary to the artificial temperature knob proposal in [30], our proposal satisfies the KMS condition for thermalization instead of an optimized approximation depending on tunable system parameters and it is promising to be scalable for large many body systems. The proposal in [32] relies on a similar idea to our proposal; its authors propose to sweep all possible energy transitions of the system with a slowly varying bath Hamiltonian strength, which can be considered as a different way of obtaining the effect of ancillae colliding to a subsystem with different bath Hamiltonian strength. Obviously, making use of only relevant transition frequencies leads to much faster thermalization and it is possible to get rid of some timescale constraints of [32] as the ancillae are supposed to be prepared in a thermal state for a time independent bath Hamiltonian before the collision in our proposal.

Our proposal can also lead to the cooling of the target system if it is possible to keep ancilla spins colder than the environment temperature. Indeed we mention two possible methods of spin cooling for the preparation of a cold environment that our scheme is well suited to. The first one is the use of frequent measurements on a two-level system interacting with a non-Markovian environment which brings the mean energy of interaction Hamiltonian to zero in order to reduce the total energy of the two-level system and its environment [33]. The application of this idea may suffer from the challenges posed by the necessary minimum frequency of the measurements. Another idea is to use quantum coherent or entangled two-level systems [34–36] to engineer the temperature of a two-level target system, which can then be used as an ancilla for the many-body system to be thermalized.

Our results can have practical significance for suggesting design principles of quantum thermalizing machines for finite many-body systems. Such devices would be compact as they can consist of few ancillae as artificial environment; they would be fast as they can engineer the target thermal state with high fidelity after a small number of collisions describing a unitary route to thermalization. These properties can be valuable for quantum thermal annealing [30] and quantum simulation applications [37], for example using superconducting circuits.

Author Contributions: Ö.E.M. and O.A. are responsible for the conceptualization, the formulation of methodology, and the theoretical analysis of the ideas introduced in the paper. S.C., Ö.E.M., and O.A. equally contributed to the discussions throughout the work, the numerical simulations, and the writing of the paper.

Funding: Steve Campbell gratefully acknowledges the Science Foundation Ireland Starting Investigator Research Grant "SpeedDemon" (No. 18/SIRG/5508) for financial support.

Acknowledgments: The authors thank Mauro Paternostro and Giacomo Guarnieri for useful discussions and feedback for early drafts of the paper.

Conflicts of Interest: The authors declare no conflict of interest.

References

1. Ford, J. The Fermi-Pasta-Ulam problem: Paradox turns discovery. *Phys. Rep.* **1992**, *213*, 271–310. [CrossRef]
2. Breuer, H.P.; Petruccione, F. *The Theory of Open Quantum Systems*; Oxford University Press: Oxford, UK, 2002.
3. Rau, J. Relaxation Phenomena in Spin and Harmonic Oscillator Systems. *Phys. Rev.* **1963**, *129*, 1880–1888. [CrossRef]
4. Scarani, V.; Ziman, M.; Štelmachovič, P.; Gisin, N.; Bužek, V. Thermalizing Quantum Machines: Dissipation and Entanglement. *Phys. Rev. Lett.* **2002**, *88*, 097905. [CrossRef] [PubMed]
5. Ciccarello, F. Collision models in quantum optics. *Quantum Meas. Quantum Metrol.* **2017**, *4*, 53–63. [CrossRef]
6. McCloskey, R.; Paternostro, M. Non-Markovianity and system-environment correlations in a microscopic collision model. *Phys. Rev. A* **2014**, *89*, 052120. [CrossRef]
7. Pezzutto, M.; Paternostro, M.; Omar, Y. Implications of non-Markovian quantum dynamics for the Landauer bound. *New J. Phys.* **2016**, *18*, 123018. [CrossRef]
8. Barra, F. The thermodynamic cost of driving quantum systems by their boundaries. *Sci. Rep.* **2015**, *5*, 14873. [CrossRef]
9. Lorenzo, S.; Ciccarello, F.; Palma, G.M. Composite quantum collision models. *Phys. Rev. A* **2017**, *96*, 032107. [CrossRef]
10. Barra, F.; Lledó, C. Stochastic thermodynamics of quantum maps with and without equilibrium. *Phys. Rev. E* **2017**, *96*, 052114. [CrossRef]
11. Chiara, G.D.; Landi, G.; Hewgill, A.; Reid, B.; Ferraro, A.; Roncaglia, A.J.; Antezza, M. Reconciliation of quantum local master equations with thermodynamics. *New J. Phys.* **2018**, *20*, 113024. [CrossRef]
12. Ziman, M.; Štelmachovič, P.; Bužek, V.; Hillery, M.; Scarani, V.; Gisin, N. Diluting quantum information: An analysis of information transfer in system-reservoir interactions. *Phys. Rev. A* **2002**, *65*, 042105. [CrossRef]
13. Cusumano, S.; Cavina, V.; Keck, M.; De Pasquale, A.; Giovannetti, V. Entropy production and asymptotic factorization via thermalization: A collisional model approach. *Phys. Rev. A* **2018**, *98*, 032119. [CrossRef]
14. Ciccarello, F.; Palma, G.M.; Giovannetti, V. Collision-model-based approach to non-Markovian quantum dynamics. *Phys. Rev. A* **2013**, *87*, 040103. [CrossRef]
15. Kretschmer, S.; Luoma, K.; Strunz, W.T. Collision model for non-Markovian quantum dynamics. *Phys. Rev. A* **2016**, *94*, 012106. [CrossRef]
16. Vacchini, B. Generalized Master Equations Leading to Completely Positive Dynamics. *Phys. Rev. Lett.* **2016**, *117*, 230401. [CrossRef]
17. Çakmak, B.; Pezzutto, M.; Paternostro, M.; Müstecaplıoğlu, Ö.E. Non-Markovianity, coherence, and system-environment correlations in a long-range collision model. *Phys. Rev. A* **2017**, *96*, 022109. [CrossRef]
18. Rybár, T.; Filippov, S.N.; Ziman, M.; Bužek, V. Simulation of indivisible qubit channels in collision models. *J. Phys. B At. Mol. Opt. Phys.* **2012**, *45*, 154006. [CrossRef]
19. Bodor, A.; Diósi, L.; Kallus, Z.; Konrad, T. Structural features of non-Markovian open quantum systems using quantum chains. *Phys. Rev. A* **2013**, *87*, 052113. [CrossRef]
20. Campbell, S.; Ciccarello, F.; Palma, G.M.; Vacchini, B. System-environment correlations and Markovian embedding of quantum non-Markovian dynamics. *Phys. Rev. A* **2018**, *98*, 012142. [CrossRef]
21. Lorenzo, S.; McCloskey, R.; Ciccarello, F.; Paternostro, M.; Palma, G.M. Landauer's Principle in Multipartite Open Quantum System Dynamics. *Phys. Rev. Lett.* **2015**, *115*, 120403. [CrossRef]
22. Karevski, D.; Platini, T. Quantum Nonequilibrium Steady States Induced by Repeated Interactions. *Phys. Rev. Lett.* **2009**, *102*, 207207. [CrossRef] [PubMed]
23. Strasberg, P.; Schaller, G.; Brandes, T.; Esposito, M. Quantum and Information Thermodynamics: A Unifying Framework Based on Repeated Interactions. *Phys. Rev. X* **2017**, *7*, 021003. [CrossRef]
24. Dağ, C.B.; Niedenzu, W.; Müstecaplıoğlu, O.E.; Kurizki, G. Multiatom Quantum Coherences in Micromasers as Fuel for Thermal and Nonthermal Machines. *Entropy* **2016**, *18*, 244. [CrossRef]
25. Seah, S.; Nimmrichter, S.; Scarani, V. Nonequilibrium dynamics with finite-time repeated interactions. *Phys. Rev. E* **2019**, *99*, 042103. [CrossRef]
26. Çakmak, B.; Campbell, S.; Vacchini, B.; Müstecaplıoğlu, O.E.; Paternostro, M. Robust multipartite entanglement generation via a collision model. *Phys. Rev. A* **2019**, *99*, 012319. [CrossRef]
27. Nigro, D. On the uniqueness of the steady-state solution of the Lindblad–Gorini–Kossakowski–Sudarshan equation. *J. Stat. Mech.* **2019**, *18*, 043202. [CrossRef]

28. Spohn, H. An algebraic condition for the approach to equilibrium of an open N-level system. *Lett. Math. Phys.* **1977**, *2*, 33–38. [CrossRef]
29. Radhakrishnan, C.; Parthasaraty, M.; Jambulingam, S.; Byrnes, T. Quantum coherence of the Heisenberg spin models with Dzyaloshinsky-Moriya interactions. *Sci. Rep.* **2017**, *7*, 13865. [CrossRef]
30. Shabani, A.; Neven, H. Artificial quantum thermal bath: Engineering temperature for a many-body quantum system. *Phys. Rev. A* **2016**, *94*, 052301. [CrossRef]
31. Delfino, G. Integrable field theory and critical phenomena: The Ising model in a magnetic field. *J. Phys. A Math. Gen.* **2004**, *37*, R45. [CrossRef]
32. Metcalf, M.; Moussa, J.E.; de Jong, W.A.; Sarovar, M. Engineered thermalization of quantum many-body systems. *arXiv* **2019**, arXiv:1909.02023.
33. Erez, N.; Gordon, G.; Nest, M.; Kurizki, G. Thermodynamic control by frequent quantum measurements. *Nature* **2008**, *452*, 724–727. [CrossRef] [PubMed]
34. Çakmak, B.; Manatuly, A.; Müstecaplıoğlu, Ö.E. Thermal production, protection, and heat exchange of quantum coherences. *Phys. Rev. A* **2017**, *96*, 032117. [CrossRef]
35. Dağ, C.B.; Niedenzu, W.; Özaydın, F.; Müstecaplıoğlu, O.; Kurizki, G. Temperature Control in Dissipative Cavities by Entangled Dimers. *J. Phys. Chem. C* **2019**, *123*, 4035–4043. [CrossRef]
36. Özaydın, F.; Dağ, C.B.; Tuncer, A.; Müstecaplıoğlu, Ö.E. Work and Heat Value of Bound Entanglement. *arXiv* **2018**, arXiv:1809.05085
37. Mostame, S.; Rebentrost, P.; Eisfeld, A.; Kerman, A.J.; Tsomokos, D.I.; Aspuru-Guzik, A. Quantum simulator of an open quantum system using superconducting qubits: Exciton transport in photosynthetic complexes. *New J. Phys.* **2012**, *14*, 105013. [CrossRef]

© 2019 by the authors. Licensee MDPI, Basel, Switzerland. This article is an open access article distributed under the terms and conditions of the Creative Commons Attribution (CC BY) license (http://creativecommons.org/licenses/by/4.0/).

MDPI
St. Alban-Anlage 66
4052 Basel
Switzerland
Tel. +41 61 683 77 34
Fax +41 61 302 89 18
www.mdpi.com

Entropy Editorial Office
E-mail: entropy@mdpi.com
www.mdpi.com/journal/entropy

www.ingramcontent.com/pod-product-compliance
Lightning Source LLC
LaVergne TN
LVHW070619100526
838202LV00012B/684